Combinatorial Materials Development

ACS SYMPOSIUM SERIES **814**

Combinatorial Materials Development

Ripudaman Malhotra, Editor
SRI International

American Chemical Society, Washington, DC

Library of Congress Cataloging-in-Publication Data

Combinatorial materials development / Ripudaman Malhotra, editor.

    p. cm.—(ACS symposium series ; 814)

  Includes bibliographical references and index.

    ISBN 0–8412–3772–7

    1. Materials—Congresses. 2. Combinatorial chemistry— Congresses.

  I. Malhotra, Ripudaman. II. Series.

TA403.6.C62 C52   2002
620.1′1—dc21   2001055315

The paper used in this publication meets the minimum requirements of American National Standard for Information Sciences—Permanence of Paper for Printed Library Materials, ANSI Z39.48–1984.

PRINTED IN THE UNITED STATES OF AMERICA

Foreword

The ACS Symposium Series was first published in 1974 to provide a mechanism for publishing symposia quickly in book form. The purpose of the series is to publish timely, comprehensive books developed from ACS sponsored symposia based on current scientific research. Occasion-ally, books are developed from symposia sponsored by other organiza-tions when the topic is of keen interest to the chemistry audience.

Before agreeing to publish a book, the proposed table of contents is reviewed for appropriate and comprehensive coverage and for interest to the audience. Some papers may be excluded to better focus the book; others may be added to provide comprehensiveness. When appropriate, overview or introductory chapters are added. Drafts of chapters are peer-reviewed prior to final acceptance or rejection, and manuscripts are prepared in camera-ready format.

As a rule, only original research papers and original review papers are included in the volumes. Verbatim reproductions of previously published papers are not accepted.

ACS Books Department

Contents

Indexes

Preface

The importance of combinatorial approaches to the development of pharmaceuticals is well recognized and can be evidenced by the fact that these methods have been embraced by every major pharmaceutical company. The combinatorial approach allows simultaneous synthesis of libraries containing thousands of compounds, a task that is unfeasible when using traditional approaches without an inordinate expenditure of resources and time. These compounds must then be screened for the desired properties. In the case of pharmaceuticals, the desired property is often the ability of the drug to bind to a receptor.

Application of the combinatorial approach to materials development is in the emerging state. Here, the properties for which the material needs to be screened are much more varied and new strategies for high-throughput screening are being developed. Nevertheless, the success of this approach for developing new optical and magnetic materials has demonstrated its viability and spawned its application to other areas such as polymers and catalysts.

A combinatorial approach provides an efficient means of screening large numbers of compounds, formulations, or even process conditions that would have been impractical by the traditional approach that uses one variable at a time. The list of target materials is large; it encompasses small organic molecules, polymer and polymer blends, ceramics, glasses, and metal alloys. The field of use is equally diverse. Examples include structural materials, magnetic, optical and photorefractive materials, catalysts, coatings, semiconductors and superconductors, adhesives, and sensors.

Combinatorial methods thus offer the possibility of shortening the time to discovery of new materials. Thus a clear potential exists for speeding up the innovative cycle-from discovery, through development, to bringing the product to market-and that is the promise of this approach. The reduced time for innovation translates into greater returns on R&D investments, and therein lies a major force driving chemical companies into adopting combinatorial approaches for their businesses, either by forming alliances with companies that are developing combinatorial tools or by acquiring the capability in-house.

As part of the Celebration of Chemistry in the 21st Century during the 219th National Meeting of the American Chemical Society (ACS) in March 2000 in San Francisco, California, the Materials Chemistry Secretariat organized a symposium on combinatorial approaches to materials development. The symposium featured presentations on four themes that are crucial to the combinatorial approach: (1) parallel synthesis, (2) rapid throughput screening, (3) robotics, and (4) informatics. The organizers of this symposium felt that the symposium should not only feature frontier research, but also should serve as a tutorial for a wider audience. Accordingly, the speakers were encouraged to provide sufficient background in these themes before discussing them. The idea for publishing this book was born during those discussions. It was felt that this collection of papers provides an overview of the field and discusses important advances that are enabling the development of new materials by combinatorial methods. This volume should provide a useful introduction to combinatorial methods for graduate students and professionals of materials science. It should be of interest to scientists actively engaged in materials research and to those considering applying this new method of materials development.

Acknowledgment

The symposium was organized under the aegis of the Material Chemistry Secretariat of the ACS, and was cosponsored by the ACS Divisions of Fuel Chemistry, Physical Chemistry, and Polymer Chemistry Inc. I am grateful for their encouragement and support in organizing the symposium. Financial support was provided by the Petroleum Research Fund of the ACS, as well as several private companies who are interested in combinatorial development of materials. In alphabetic order, they are: Agilent Technologies, Argonaut Technologies, Bohdman Automation, Genevac Technologies, Symyx Technologies, Tripos Inc., and VWR Scientific. I am most grateful to Chery Lund, the coorganizer of the symposium, for her tireless efforts in putting together a stellar line of speakers. The authors and reviewers of the chapters deserve special thanks for their commitment to this project. I am grateful to Robert Haushalter, who was most generous with his time and helped me vet some of the papers. I am also grateful to Kelly Dennis and Stacy Vanderwall in acquisitions and Margaret Brown in production of the ACS Books Department.

Ripudaman Malhotra
SRI International
333 Ravenswood Avenue
Menlo Park, CA 94025
Phone: (650) 859–2805
Fax: (650) 859–6196
Email: ripu@sri.com

Chapter 1

Combinatorial Chemistry of Materials, Polymers, and Catalysts

Wilhelm F. Maier, Guido Kirsten, Matthias Orschel, Pierre-A. Weiss, A. Holzwarth, and Jens Klein

Max-Planck-Institut für Kohlenforschung, Kaiser-Wilhelm-Platz 1, D—45470 Mülheim an der Ruhr, Germany

Our standard of living relies on the discovery and availability of materials (solids with a function). The fulfillment of many desires, such as environmentally friendly chemical production, moisture transporting fabrics, brilliant and stable colors, water- or scratch resistant paints, faster computers, smaller cellular phones, low energy cars or light bulbs and many others depends primarily on the discovery and development of suitable materials. Unfortunately, the discovery of new materials or catalysts is still dominated by serendipity, chemical intuition and systematic rows of experiments. Our chemical knowledge does not allow to predict novel chemical compositions or structures for materials based on desired properties. Better materials are searched for by conventional chemical research, based on one experiment at a time and trial and error. State of the art chemistry is well structured for the optimization and improvement of special properties of known materials. However, it does not provide a concept for the dedicated search for new or novel materials. Analogy, derivatisation and modification dominate conventional research in chemistry. The demand for new or better materials is unlimited and cannot be satisfied by conventional methods. Combinatorial chemistry is the first rational concept to tap the potential wealth of chemical diversity in a more systematic way. The original meaning of combinatorial chemistry was the creation of molecular diversity by the synthetic combination of molecular fragments or building blocks. Today combinatorial chemistry stands for a rapidly increasing number of methods for the preparation, characterization and screening of libraries of molecules or materials. An excellent and detailed review about the state of the art of combinatorial

1

chemistry in materials research, covering mainly 1995-1999, has just been published (1). Due to the large number of elements and an almost infinite number of building blocks the possible diversity of chemistry rapidly reaches astronomical dimensions never accessible to systematic or comprehensive surveys. The huge potential for new discoveries and new lead structures in chemistry can be elucidated with a few numbers. Bohacek et al.(2) have estimated from structural consideration, how many potentially stable organic structures are possible, based on the elements C, O, N, S (and H). They came to the astronomical estimate of 10^{63} stable structures for molecules based on up to 30 of the above elements. Preparing only 1 mg each a total mass of 10^{60} g would be obtained. The total mass of the earth is ~ $5*10^{27}$ g, that of our sun ~ $2*10^{33}$ g, which means that the total mass of this organic library is equivalent to the mass of 10^{27} suns! The mass of the whole known universe is estimated to 10^{58} g. When we generously assume a total number of 20 million of presently known organic compounds, with 1 mg each a total mass of just 20 kg represents our present "chemical universe", negligible when compared with 100 times the mass of the universe. Even with combinatorial methods this diversity can never be explored comprehensively, these methods can only help to find new lead structure more often than in the past. Even more problematic becomes reality, when we move from organic diversity of only five elements to the diversity of the periodic table. Assuming 75 useful and stable elements a compositional diversity (excluding different orders) of about 2800 binaries (one of it is CH = all hydrocarbons!), $7*10^4$ ternaries, $1.3*10^6$ quaternaries and already over $6*10^{12}$ decanaries are possible, and this without even considering stoichiometric and structural diversity. The structural and stoichiometric diversity, potentially contained in every individual combination of elements, therefore brings a second astronomical dimension in this simple minded view of chemical diversity. Since the total number of known inorganic structures is far smaller compared to organic, it can be concluded that the little, what is known in chemistry, becomes negligible compared with the possibilities offered by the periodic system. Combinatorial chemistry is presently the only rational concept offering a more effective access to this diversity, whereby it seems obvious, that a systematic and comprehensive access to chemical diversity will be impossible. Combinatorial chemistry tackles the potential diversity of chemistry in a rational fashion. Instead of sequential one at a time experiments, a large number of new or known components are synthesized in parallel and investigated in combinatorial libraries. High throughput screening methods are used to rapidly search large libraries for desired properties. However, reliable combinatorial techniques for the discovery and development of new materials are by no means trivial. The preparation of libraries has to be automated and suitable protocols have to be developed. Mixing, grinding and heating of fine powders, hydrothermal syntheses, polymerization, sol-gel-procedures or precipitation reactions are classic methods for the preparation of solids. Since no structural synthetic concept exists (in contrast to organic synthesis), the combinatorial preparation of materials libraries can easily lead to the formation of novel materials, who's structure and properties depend on the preparation and calcination procedures applied. To establish a combinatorial chemistry of materials therefore requires reliable library preparation, component characterizations, property assays and a structural database to manage the information flow. Motivation for the preparation of materials libraries is the discovery of new solids with a desired property profile. There is no limitation for properties suitable for combinatorial searches: electric properties (such as conducting, dielectric, magnetic, superconducting, isolating, electrochromic), materials properties (such as scratch resistant, elastic, hard, temperature stable, light, corrosion resistant, UV-resistant), catalytic properties (such as basic, acidic, porous, shape selective, oxidizing, reductive, oxygen activating, hydrogen activating, isomerizing), sensoric properties (such as highly selective response to selected gases or molecules, molecular recognition), special memory properties, optical properties (such as colors, optical switching, frequency doubling, lasing) and many others. Demand will determine the properties, which will become subject of combinatorial developments. Prerequisite is the availability of suitable high throughput assays. The first documented combinatorial approach (then called "Multiple Sample Concept") to materials research was published already in 1970 by Hanak, who used gradient sputtering techniques to find new superconducting materials (binary metal alloys). In the beginning of this publication he already questions the efficiency of conventional research: Hanak was clearly ahead of time, because it took another 25 years before Schultz published the next superconductor library consisting of an array of 128 single components in 1995.(3) This publication marks the beginning of combinatorial chemistry for distinct materials. Schultz et al. applied an elegant and rarely used synthesis technique based on the sequential deposition of thin films (thickness in the range of sub-monolayer to few atomic layers) by controlled Johnson et al. have advanced this technique(14) and with it many new solid phases have been identified already. The principle of this method is repeated sequential deposition, producing layered thin films with thicknesses in the range of monolayers. Almost any stoichiometry can be obtained. Due to low diffusion barriers between very thin layers homogeneous amorphous mixtures or novel kinetically stable phases form already at relatively low temperatures. The advantage of this procedure is, that it can access most elements of the periodic system as synthetic components and in combination with state of the art masking techniques and temperature treatments, libraries of almost unlimited size and composition can be prepared. In 1997 SYMYX-technologies published a luminescent materials library containing 25.000 components on a small wafer, which mainly demonstrates the state of the art of combinatorial library preparation. In such studies a novel luminescent material, $SrCeO_4$ (blue), has been discovered. A potential disadvantage of component preparation through thin film annealing is the lack of associated laboratory procedures to reproduce interesting components in sufficient quantity. Another attractive technique is the use of ink-jet technology for library preparation under liquid phase preparation conditions, first reported by Schultz et al.(5) Here the ink-solutions are simply replaced by chemical solutions and the high spatial resolution together with the precise dosing of nL-quantities can be utilized conveniently for liquid phase synthesis of minute amounts of materials. Ink-jet technology has also been used for the library preparation in a search for better catalysts for the methanol fuel cell. One of the limiting problems of present fuel cell technology is the low current density of the anode catalysts. On conducting Toray carbon paper as library support material, with 220 compositions the phase diagrams of ternary and quaternary alloys of Pt, Rh, Os, Ru and Ir have been mapped out and the half cell performance of the individual compositions were identified by the use of a pH-sensitive fluorescence indicator. In this study new fuel cell catalysts have been discovered ($Pt_{62}Rh_{25}Os_{13}$ and $Pt_{44}Ru_{41}Os_{10}Ir_5$), which provided almost 50 % higher current densities than the well established $Pt_{50}Ru_{50}$ catalyst used traditionally.(6) In the meantime several laboratories have already reported library preparations with more conventional liquid phase preparation conditions based on the use of synthesis or pipetting robots. We are experienced in the use of sol-gel methods for catalyst preparation.

Figure 1: synthesis robot

In Figure 1 our synthesis robot is shown, the rack with precursor solutions used by the robot to compose the preparation solutions for each catalyst is shown in figure 2.

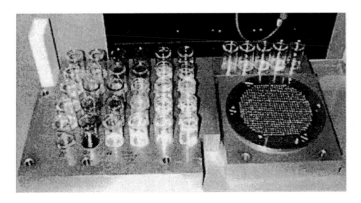

Figure 2: pipetting robot configuration with library and precursor solutions

A picture of a ready made library (diameter 10 cm) containing 760 potential catalyst materials, seen through the transparent window of a reactor is shown in figure 3.

Figure 3: Gas phase reactor with library under IR transparent window

The library material is slate, the diameter of the individual component holes is 2 mm containing about 2-5 mg of material each. The preparation time for such a library is about 2 h, while 1-2 days are needed to write the synthesis program and prepare all the stock solutions. While the libraries mentioned above rely on synthesis protocols acceptable for robots and secondary treatments are limited to normal pressure. Many inorganic materials, especially oxides, require extreme reactions conditions, such as hydrothermal ones. These procedures are often very time consuming and cumbersome and could thus benefit greatly from combinatorial approaches. However, especially due to the high pressures and temperatures involved, combinatorial procedures are not straight forward. Wendelbo et al. have first published the application of combinatorial methods for hydrothermal synthesis.(7) Using an array of 100 1ml autoclaves 100 hydrothermal syntheses have been carried out in parallel at temperatures up to 200°C. The phase diagram obtained in such a single experiment demonstrates the potential strengths of combinatorial approaches in the area of the synthesis of solid state materials. In these experiments, reaction volumes were still in the range of 0.5 mL. The identification of the phases prepared required still manual removal of the solids from the reactors followed by individual measurements with conventional X-ray diffraction techniques, the required sample amount is in the range of 1-10 mg. The sample handling after synthesis is still quite cumbersome. Every sample has to be prepared individually. In a parallel study (at the same time) we have been engaged in the development of a technique, which allows the preparation of inorganic materials on a library under pressure and high temperature. Emphasis was not only paid to a highly parallelized synthesis method, but also to the preparation of a library, which can be analyzed directly by suitable high throughput techniques.(8)

The small reactor is sandwiched in a steel housing. The synthesis reactors are small drill holes in a Teflon plate (reactor walls), while the reactor bottom (library) is a Si-single crystal wafer. On top of the Teflon plate is a steel mask, which presses the Teflon plate against the Si-wafer with the help of screws. Through the holes of the steel plate (same drilling pattern as the Teflon plate) the reaction solutions are introduced (1-2 µL total reaction volume). The top of the steel mask is sealed by another Teflon plate, which itself is screwed down by the top steel plate. As reaction solutions a TS-1synthesis was modified by the use of different templates and concentrations, introducing a total of 37 pre-prepared solutions with an Eppendorf automated dosing pipette. The reactor was closed and heated in a simple furnace for 36 h at 200°C. The products were then washed carefully on the library with distilled water and then dried and calcined. After calcination the completed library is ready for X-ray analysis. The X-ray powder diffraction pattern have been obtained automatically by the commercially available GADDS microdiffractometer. The instrument allows spot analysis of areas as small as 50 µm. The library is mounted on the XYZ-table and the controlling computer allows convenient teaching of the library spots to be analyzed. A 1 min measuring time per spot was found sufficient.(8)

The experiment has shown, that hydrothermal synthesis can be carried out on a 2 µL-scale in highly parallelized fashion for the direct preparation of libraries. The next step was to increase automation and library size (10 cm diameter). With increasing size leakage between reaction chambers increased and especially with highly basic solutions corrosion problems increased. Figure 4 shows our present reactor for hydrothermal synthesis, which can be filled automatically by a pipetting robot, but development is not finished and at present there is nothing new to report.

Figure 4: next generation of hydrothermal reactor

A semiautomatic combinatorial zeolite synthesis has been reported just recently by Bein et al. (9). In a more conventional approach autoclave arrays have been filled by pipetting robots, products have been isolated by automated filtration and centrifuging, but product analysis by XRD is carried out conventionally.

In general, reliable high throughput detection of activity has emerged as one of the limiting parameters of combinatorial applications in catalyst development. A very attractive method to simultaneously monitor the activity of catalysts on a library is IR-thermography. Studying the highly exothermic formation of water from hydrogen and air with impregnated pellets of catalysts (arranged on a flat plate in a reactor) Wilson et al.(10) have shown, that IR-thermography can differentiate among active and inactive catalysts. Here temperature differences around 60°C are recorded between library and active pellet. More temperature sensitive is the method reported by Taylor and Morken(11), who identified active acyl transfer catalysts by the heat of reaction with an IR-video camera in a pool of 7000 encoded beads containing 3150 discrete compounds, floating on the surface of the $HCCl_3$-solution. The most active beads were identified by small temperature differences of about 1 K. This observation of absolute heat differences works well for high temperature differences. However, for less exothermic reactions and especially for smaller amounts of catalysts a more sensitive method was required.

Figure 3 shows a library (made of slate) in our reactor, it can be seen through an IR-transparent window (sapphire). The library consists of 760 AMM-catalysts (amorphous microporous mixed oxides, made by sol-gel synthesis) of about 200 µg each.

Several problems had to be overcome. In contrast to visible imaging, where a light source is required and the image is composed exclusively of the reflected light, IR imaging at normal or high temperatures is composed of active IR-emission, IR-reflection and temperature differences. The IR emission of molecules or surfaces is dependent on emissivity, a material specific property. Reflection is dependent on an IR-radiation source and the material specific reflection properties and temperature differences result in radiation differences following the laws of black body radiation. A library was prepared by pipetting precursor solutions for the preparation of AMM-type catalyst into the 2 mm holes, 2mm deep, present in a library substrate made from slate. Amorphous microporous mixed oxides (AMM) are promising new materials, whose catalytic properties can be controlled during a one-step preparation procedure of the sol-gel-type. By acid catalyzed co-polycondensation of an alkoxide of Si, Zr, Al or Ti with a soluble derivative of a catalytically active element (such as Mn, Mo, V, Ti, Sn, In, Cu, Fe, Cr.....), atomically distributed active centers in the oxide matrix can be obtained. In the absence of any template, the sol-gel reaction, followed by drying and calcination, provides a highly porous mixed oxide with a narrow pore size distribution around 0.6-0.9 nm. The surface polarity of such an oxide can be controlled by copolycondensation with methyltriethoxysilane, The basic reaction for silica based AMM-materials is shown below:

$$xM(OR)_n + yMeSi(OR)_3 + (RO)_4Si + H_2O \rightarrow (MO_{n/2})_x(Me-Si1_{1/2})ySiO_2 + ROH$$

catalytic	polarity	microporous	Amorphous Microporous
centres	modifier	matrix	Mixed oxide (AMM)

The materials are denoted as AMM-M_xM', where M' stands for the base oxide and x for the atom% of the additional oxide M (f.e. AMM-Ti_3Si = 3 % titania in 97 % silica). After proper calcination, these materials remain amorphous and have a porosity of 10-35%. The preparation procedure is well suited for pipetting robots.

In Figure 5 the IR-image of a catalyst library at constant temperature is clearly visible, not due to temperature differences, but due to emissivity and reflectivity differences of the different materials, as mentioned above.

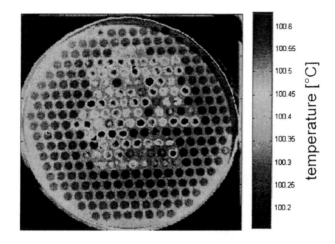

Figure 5:IR-image without emissivity correction

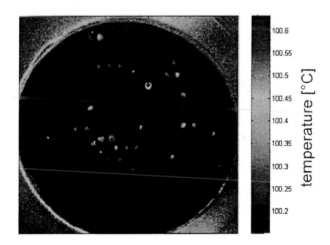

Figure 6 IR-image

With the small catalyst amounts (< 1 mg) required for large libraries the catalyst components are present as thin films, which tend to dissipate small temperature increases rapidly into the bulk library substrate. This requires a method sensitive to the smallest temperature changes. Clearly, if temperature differences are smaller than or similar to the emissivity differences shown in Fig. 5 there is a problem. For the purpose of imaging heats of reactions by small temperature differences it is necessary to separate the pure temperature differences from emissivity and reflexivity differences. This was achieved by our development of emissivity corrected IR-imaging. Here the IR image of the reactor at the reaction temperature is recorded and used as basic correction, from which the actual IR-images are subtracted in real time and then displayed. The method(12) takes advantage of the fact, that as long as the catalyst does not change, its emissivity remains stable during the reaction. After proper emissivity correction the recorded image changes into a featureless blank screen without any contrast.

Now any small change in temperature is identified easily.

Figure 6 shows heats of reaction from the oxidation of propene with air at 100°C. Note the small temperature differences recorded and that only few of the catalyst spots are active.

Figure 7 shows the IR-thermographic image of the large AMM-library from Fig. 3 during the oxidation of methane with air to carbondioxide at 400°C.

The library is extremely divers, it contains about 3 mol% each of 70 different catalytically active centres dispersed in hydrophilic and hydrophobic micro

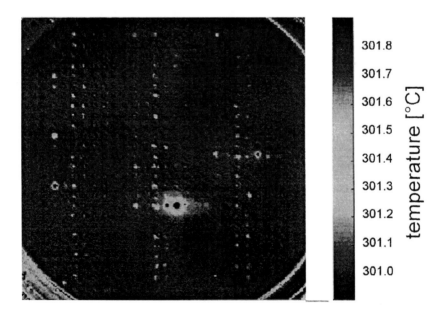

Figure 7:IR-image

Figure 8: IR-setup

porous oxides of Si, Al, Ti, and Zr. It can be clearly seen, that many materials are not active, and some show quite different activity. It should also be noted, that the observed temperature range (shown as colour scale at the right) is very narrow with resolutions better 0.1°C.

In Figure 8 a photograph of the principal set-up of the IR-thermography in the laboratory is shown.

Below the camera the IR-reactor with supply capillaries and preheater can be seen. At the right the computer for image-monitoring, a quadrupole MS for monitoring integral product gas composition, an oscilloscope for camera optimization at reaction conditions and several temperature and mass flow controllers can be identified. Clearly, it would be beneficial, if the relative total heat of reaction could be determined from the IR-image to allow a more precise scaling of the various active materials identified

A different approach has already been used successfully by SYMYX, who reported new ethane dehydrogenation catalysts based on optimized library development.(13) For activity detection and comparison a special laser based technique had to be developed. Primary screening went through about 10.000 catalyst compositions per month. Selectivities of 88 % at 10 % conversion are reported for the optimized oxide catalysts (typical composition $Mo_{71}V_{24}Nb_{0.5}Sb_2Li_{2.5}O_x$)(7). Emissivity corrected IR-thermography has the advantage of speed and direct visualisation of complex data. Of disadvantage is the increase of signal intensity with combustion. If selective oxidation is subject of interest, IR-thermography may be used for primary screening, but another method has to be applied for product identification at each library position. Very attractive is spatially resolved mass spectrometry. Problematic is the spatial adressing of the library spots, especially with 2-dimensional libraries. Possible options are capillary bundles, where each library spot is addressed by the end of an individual capillary, and the mass-spectrometer scans through all capillaries in a sequential fashion. Dead time and analysis time linearly add to the total measuring time. An intriguing option to us was here the direct use of the catalyst library as an array of micro reactors. A very simple protocol for the spatially resolved detection of catalytic activity and selectivity by mass spectrometry results from the coupling of a synthesis robot with a commercial gas analysis unit with capillary inlet. Of advantage here is, that the synthesis robot knows already the position of the library spots. Figure 9 shows the principal arrangement of our first successful screening for catalytic selectivity in the oxidation of propene with air.

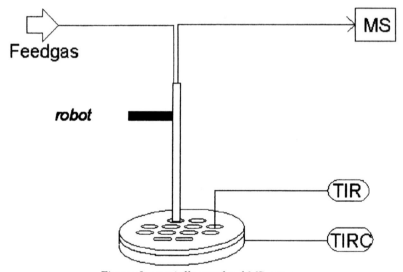

Figure 9: spatially resolved MS-setup

The pipetting unit of the robot is exchanged by a capillary bundle holding the capillary of the inlet system of a conventional gas analyser (quadrupole MS) and the feed gas capillary (0.2 mm diameter). The scheme in Figure 10 illustrates the measuring principle and the conversion of the library into an array of micro reactors.

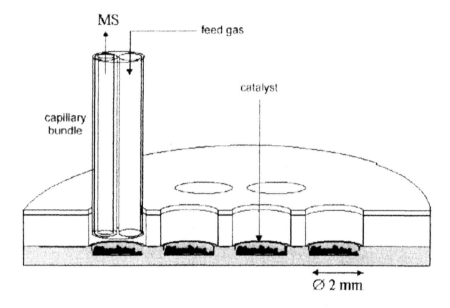

Figure 10: catalyst screening on a library with a capillary bundle

The library is heated directly from below through an electric heating plate. Radiative losses above the library are reduced by an additional ceramic mask placed on top of the library. Its hole pattern matches that of the library thus enlarging the "reactor volume" above the catalyst components. The capillary bundle is inserted sequentially into each library compartment thus shielding the reactants from the laboratory air. Unusual is the open construction, where the feed gas flow is directed onto the catalyst and the mass spectrometer samples and analyses a mixture of feed gas and products. The excess feed gas / product mixture, which effectively displaces the air above the library component, leaves the sampled library compartment through the remaining space between the capillary bundle and the drill hole wall of the ceramic mask. On top of the ceramic mask an additional brass plate with conical holes helps to guide the capillary bundle into the library compartments.(14)

It has been shown, that the screening results obtained for a library of 600 catalysts provide a qualitatively correct information on the different activities of the catalysts studied. Besides CO_2 as undesired product of total oxidation the following selective oxidation products have been observed: acrolein, 1,5-hexadiene, benzene, allyl alcohol and propylene oxide. The selectivity for propene oxidation of several AMM-type catalysts, observed in the gas phase under conventional reaction conditions(15) served as a reference.

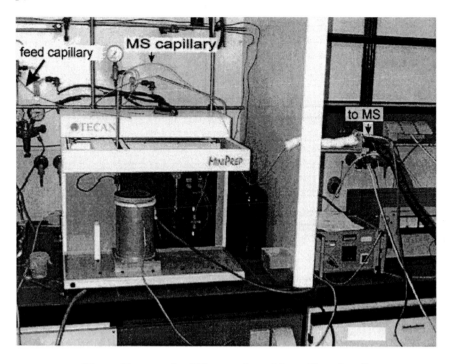

Figure 11: setup for MS-screening with capillary bundle

One serious problem encountered due to the small total amount of catalyst used is selectivity. Catalysts prepared by a sol-gel procedure with the synthesis robot on a library often form solid glass bodies (thin film catalysts), which have very little outer surface. On such catalysts the sensitivity of the quadrupole is too low to detect any product formation. A possible solution is the use of powdered catalysts on the library, which due to the macroporosity of packed powder produce enough product to be detected readily by a quadrupole. However, the advantage of automated synthesis is lost. By replacing the quadrupole MS with a double focusing high resolution mass spectrometer enough sensitivity is gained to detect the minute amount of product formed on the small outer surface (one or a few solid particles) of catalysts in robot synthesized libraries. The photograph in Fig. 11 shows the robot with the MS-interface and the double focusing mass spectrometer used. This experiment also illustrated the flexibility of this robot based technique, where the analysis unit can be exchanged rapidly.

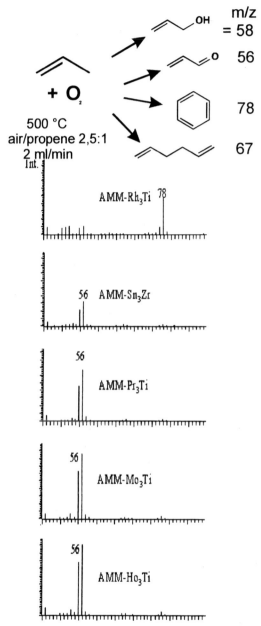

Figure 12: Results obtained by double focusing MS

In Figure 12 the results obtained for the selective oxidation of propene with air on thin film catalysts in the robot generated library are presented. The mass spectra shown were selected from the complete scan of the library (600 catalysts), obtained at 500°C. The catalysts selected are all new catalysts. There are no reports in the literature on such materials for selective oxidation of hydrocarbons. Masses 54 and 56 are typical for acrolein formation, while mass 78 indicates benzene formation. Figure 13 therefore shows true discoveries by primary screening techniques, although confirmation under conventional conditions has not been carried out yet. It should be emphasized, that none of these catalysts showed any detectable product formation with the regular quadrupole MS under identical reaction conditions.

Another advantage of the double focusing MS is the high resolution such instruments allow. Compounds of identical mass, but different chemical composition, such as ethylene, nitrogen and CO (m/z = 28) cannot be differentiated by a quadrupole, but readily by high resolution mass spectrometers. This additional feature can become very useful during the high throughput screening of catalytic reactions.

Figure 13: double focusing MS

In the meantime there have been several reports on various methods of spatially resolved mass spectrometry for application as high throughput analysis (HTA) method with libraries. SYMYX has developed a dedicated quadrupole mass spectrometer, especially developed for high throughput screening during combinatorial catalyst optimization, which allows rapid screening of libraries under reaction conditions covering a broad range of temperature and pressure(16). Groups in Germany and in the US have presented techniques, where a rotating capillary interfaces a quadrupole MS with catalyst tube arrays(17,18). ESI-MS has been used successfully by Chen for HTA of polymers and oligomers(19). Even enantioselectivity, traditional domain of chromatographic methods with their associated time delay, has become accessible to HTA with ESI-MS by the intelligent use of pseudo racemates as starting materials (20). In contrast to organic libraries, microstructure and composition of the components of solid state libraries are a priori unknown, and are, as already mentioned above, usually an unknown function of preparation conditions and library treatment. Suitable techniques for spatially resolved characterization of library components are therefore of equal importance for this chemistry as library preparation or the development of suitable assays for the detection of desired materials properties. Simple optical inspection for library damage is obvious. Very attractive are nondestructive techniques, such as X-ray based methods. Spatially resolved microdiffraction with commercially available fully automated instruments allows the automatic accumulation of diffractograms of all library components, providing important structural information. Chemical composition of the library components can be identified rapidly and automatically by spatially resolved micro-X-ray fluorescence. We have been using energy dispersive X-ray fluorescence to confirm component composition in a routine manner. Compared to wavelength dispersive XRF, EDX allows qualitative and quantitative analyses in a short time with sufficient accuracy. With the commercial instrument EAGLE II from EDAX a spatial resolution of 100 µm can be obtained routinely by the use of a capillary lens. Chemical composition of small library spots as well as compositional homogeneity of the prepared materials can be monitored. Typical time required for a single spot analysis is in the range of 1 min. Figure 14 shows the principle technique.

The instrument itself is a compact table top instrument. The 2 libraries shown in figure 15 and 16 have been analyzed automatically. Despite the problems of background activity of the slate library (Fe, Si, Al) the chemical analysis results shown on Fig. 17, obtained from a single component (0.5 mg) are in perfect agreement with the expected composition according to the synthesis receipt.

The last figure (Figure 18) illustrates the largest problem. While we and others have assembled several high throughput methods for library preparation, component characterization and property identification, there is no working database to handle the amount of data generated in a transparent and efficient manner. These database programs must be able to handle preparation, characterization and property, interface all computer controlled processes involved in combinatorial projects and must be capable of projections based on the data accumulated. Such databases must be involved not only in data retrieval and evaluation, but also in library design of higher library generations.

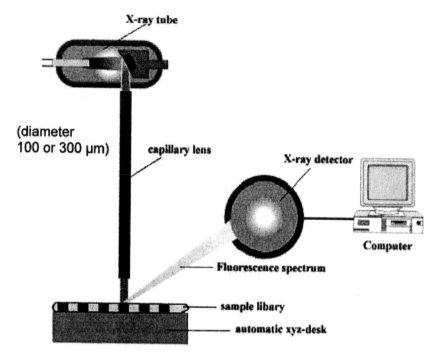

Figure 14: X-ray fluorescence setup

Figure 15: catalyst libary on PMMA-plate

Figure 16: catalyst library on slate plate

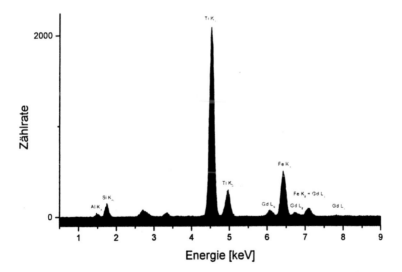

Figure 17: results

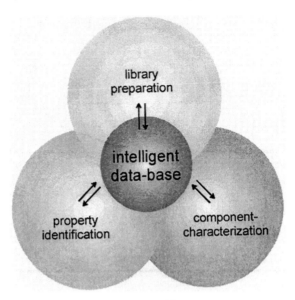

Figure 18

In summary it can be concluded, that within a few years combinatorial chemistry of materials has emerged into an important new research area. Impressive high throughput techniques have already been developed and the first discoveries due combinatorial approaches have been reported. Germany, the European Community, the US and Japan have started dedicated funding programs to support the development of this new branch of chemistry in industry and academia. Especially in academia the potential of combinatorial research should concentrate on the discovery aspect and the invention or initiation of suitable combinatorial methods in all areas of materials and catalyst research. We are convinced, that in the close future we will observe a rapid growth of combinatorial chemistry with innovative new techniques as well as new material discoveries.

Acknowledgments
The text in this chapter is based on material presented at the UNESCO Conference on Macromolecules & Materials Science on April 10-12, 2000 at Stellenbosch, South Africa

References:
(1): B. Jandeleit, D.J. Schaefer, T.S. Powers, H.W. Turner, W.H. Weinberg, Angew. Chem. Int. Ed. 38 (1999) 2494
(2): R. S. Bohacek, C. McMartin, G. Colin, C. Wayne, Med. Res. Rev. 16 (1996) 3
(3): G. Briceño, H. Chang, X. Sun, P. G. Schultz, X.-D. Xiang, Science 270 (1995) 273
(4): D.C. Johnson, Chem. Mater. 81 (1998) 3998
(5): X.-D. Sun, K-A. Wang, Y. Yoo, W. G. Wallace-Freedman, C. Gao, X.-D. Xiang, P. G. Schultz, Adv. Mater. 9 (1997) 1046.
(6): E. Reddington, A. Sapienza, B. Gurau, R. Viswanathan, S. Sarangapani, E. S. Smotkin, T.E. Mallouk, Science, 280 (1998) 1735.
(7): D. E. Akporiaye, I. M. Dahl, A. Karlsson, R. Wendelbo, Angew. Chem., 110 (1998) 629
(8): J. Klein, C. W. Lehmann, H.-W. Schmidt, W. F. Maier, Angew. Chem. Int. Ed. Engl. 37 (1998) 609
(9): T.Bein et al.,Angew. Chem. 38 (1999) 1891
(10): F. C. Moates, M. Somani, J. Annamalai, J. T. Richardson, D. Luss, R. C. Willson, Ind. Eng. Chem. Res. 35 (1996) 4801
(11): S. J. Taylor, J. P. Morken "Thermographic Selection of Effective Catalysts from an Encoded Polymer-Bound Library", Science 280 (1998) 267
(12): A. Holzwarth, H.-W. Schmidt, W.F. Maier, Angew. Chem. Int. Ed. 37 (1998) 2644.
(13): P. Cong, A. Dehestani, R. Doolen, D.M. Giaquinta, S. Guan, V. Markov, D. Poojary, K. Self, H. Turner, W.H. Weinberg, Proc. Natl. Acad. Sci. USA 96 (1999) 11077
(14): M. Orschel, J. Klein, H. W. Schmidt, W. F. Maier, Angew. Chem. 111 (1999) 2961
(15): H. Orzesek, R. P. Schulz, U. Dingerdissen, W. F. Maier, Chem. Eng. Tech. 22 (1999) 691
(16): P. Cong, R.D. Doolen, Q. Fan, D.M. Giaquinta, S. Guan, E.W. McFarland, D.M. Poojary, K. Self, H.W. Turner, W.H. Weinberg, Angew. Chem. Int. Ed. 38 (1999) 484
(17): U. Rodemerck, P. Ignaszewski, M. Lucas, P. Claus, Chem. Ing. Tech. 71 (1999) 873
(18): S. Senkan, K. Krantz, S. Ozturk, I. Onal, Angew. Chem. Int. Ed. 38 (1999) 2794
(19): C. Hinderling, P. Chen, Angew. Chem. Int. Ed. 38 (1999) 2253
(20): M.T. Reetz, M.H. Becker, H.-W. Klein, D. Stöckigt, Angew. Chem. Int. Ed. 38 (1999) 1758

Chapter 2

Combinatorial Polymer Science: Synthesis and Characterization

J. Carson Meredith[1], Archie P. Smith[2], Alamgir Karim[2],
and Eric J. Amis[2]

[1]School of Chemical Engineering, Georgia Institute of Technology, 778
Atlantic Drive, Atlanta, GA 30332–0100
[2]Polymers Division, National Institute of Standards and Technology, 100
Bureau Drive, Stop 8542, Gaithersburg, MD 20899–8542

Combinatorial methodologies offer the ability to efficiently measure relevant chemical and physical properties over large regimes of variable space. We review recent work in the development of combinatorial methods for polymer synthesis and characterization. The combinatorial synthesis of polymers including fluorescent sensors, molecular imprint systems, dendrimers, and biodegradable polymers is presented. The fundamental characterization of polymer films and coatings with combinatorial methods is also discussed. We illustrate four new techniques for preparing continuous gradient polymer libraries with controlled variations in temperature, composition, thickness, and substrate surface energy. These libraries are used to characterize fundamental properties including polymer blend phase behavior, thin-film dewetting, block copolymer order-disorder transitions.

23

Introduction

Fundamental research of the synthesis and characterization of polymeric materials is driven by their use in applications including structural materials, packaging, microelectronics, coatings, biomedical materials, and nanotechnology. Current trends demand finer control of chemistry, morphology, and surface topography at the micrometer and nanometer scales. To achieve these goals, there are increasing needs for the synthesis and processing of multicomponent mixtures, composites, and thin films. However, these systems are inherently complex due to the interactions of phase transitions, microstructure, interfaces, and transport behavior that occur during synthesis and processing. The synthesis of polymers by emulsion polymerization, for example, involves colloid chemistry, micellization, transport between phases, and complex rate relationships. In the case of coatings and thin films, the mechanical and optical properties, microstructure, and phase and wetting behavior are sensitive and poorly understood functions of thickness. In addition to the complex phenomena involved in polymer synthesis and processing, there is a large variable space involving parameters whose effects often counteract one another. These include reactant composition and structure, synthetic sequence, solvent, temperature, annealing history, pressure, and thickness (e.g., in films). Conventional microscopy, spectroscopy, and analytical tools for polymer synthesis and characterization were designed for one-sample-one-measurement utilization, and are suited for detailed characterization over a limited set of variable combinations. This conventional approach is preferred when the most relevant variable combinations are known *a priori* or can be reliably predicted from theory. However, the complex phenomena and large variable spaces present in multicomponent, multiphase, bioactive or thin film polymers often exceed the capabilities of current theory and conventional measurements. Therefore, a strong need exists for experimental techniques capable of highly efficient synthesis and characterization of complex polymeric systems over large numbers of variable combinations.

Combinatorial methods (CM) use experimental design, library creation, high-throughput screening, and informatics to efficiently and rapidly develop new materials and measure properties over large numbers of variable combinations, Figure 1. This is accomplished by preparing samples not one at a time, but rather as sample 'libraries' containing hundreds to thousands of variable combinations each. High-throughput measurements of relevant chemical and physical properties, combined with informatic data analysis, allow efficient development of structure-processing-property relationships. The benefits include efficient characterization of novel regimes of thermodynamic

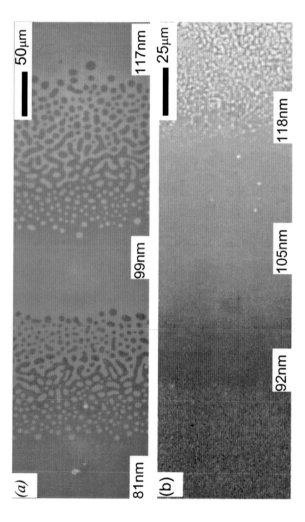

Plate 1. (a) Optical micrograph of a $M_w = 26,000$ g/mol PS-b-PMMA gradient film showing the addition of two lamellae to the surface. Labels indicate h = 4.5 L_o, 5.5 L_o and 6.5 L_o for this M_r copolymer. b) Optical micrograph of a $M_w = 104,000$ g/mol PS-b-PMMA gradient film showing the change in h across the smooth region. The color change from purple to blue-green indicates a Δh of $\approx$ 25 nm across the smooth area. Standard uncertainty in thickness is $\pm$ 3 nm.

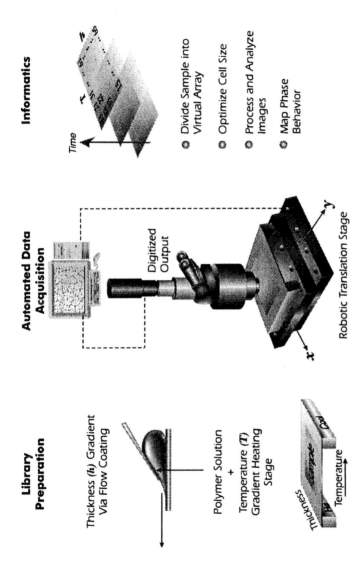

Figure 1. Schematic of the combinatorial experimental method, as applied to the preparation of thickness and temperature gradient film libraries, high-throughput screening with optical microscopy, and informatic analysis of image data as a function of temperature, thickness, time. Adapted with permission from ref. (22).

and kinetic behavior (knowledge discovery) and accelerated development of functional materials (materials synthesis and discovery). Although historically applied to pharmaceutical research, there is increasing interest in applying CM to materials science, as indicated by recent reports of combinatorial methodologies for a wide range of inorganic (*1-8*) and organic / polymeric materials (*7,9-24*).

Early combinatorial *materials* research used sputtering methods to prepare composition-gradient libraries for measuring the phase behavior of ternary metal alloys (*20*) and other inorganic materials (*25*). However, limitations in computing capacity and instrument automation overshadowed the benefits of combinatorial materials characterization until only recently. The primary limitation to *characterizing* polymers, as bulk material or films, with combinatorial methods has been a shortage of techniques for preparing libraries with systematically varied composition (ϕ), morphology, physical form, thickness (h), and temperature (T). Here, we briefly review recent advances in applying CM to polymer synthesis and characterization. We first present applications of CM to the synthesis of a wide range of polymeric materials, including sensors, dendrimers, and biodegradable polymers. Next we turn our attention to several novel methods developed by the authors for the preparation of T, ϕ, h, and surface energy continuous polymer film libraries. We focus in particular on the novel library preparation and high-throughput screening steps, since these have been the principal limiting factors in CM development for polymers. The use of continuous gradient libraries in the measurement of fundamental properties is described for polymer blend phase behavior, block copolymer segregation, and dewetting transitions.

Combinatorial Polymer Synthesis

Despite the recent increase in combinatorial inorganic materials research (*2-7,14,15*), there are still relatively few studies reporting CM for the synthesis of polymeric materials. One example is the work of Brocchini *et al.* (*10,11*), where a 112 member combinatorial library of biodegradable polyarylates was prepared by copolymerizing all possible combinations of fourteen tyrosine based diphenols and eight diacids. The pendant chain and backbone structures were systematically varied by the addition of methylene groups, substitution of oxygen for the methylene, and addition of branched or aromatic structures. The library products displayed diverse properties, as indicated by measurements of the glass transition temperature (T_g), air-water contact angle (θ_w), and fibroblast proliferation during cell culture. Fibroblast proliferation was found to decrease with increased hydrophobicity except for the main-chain oxygen containing polymers that served as uniformly good growth substrates regardless of the hydrophobicity.

The results of the above study were utilized by Reynolds (*13*) to test informatic methods for designing *diverse* and *focused* combinatorial libraries. Molecular topology and genetic-algorithm-optimized quantitative structure-property relationships were used to design libraries. These techniques allowed selection of a representative subset of library members for rapid study of the entire library (a diverse subset) or concentration on a specific property of interest (a focused subset). Each monomer pair of the 112 member library was represented by a 2-D topological descriptor, used by the algorithm to select a structurally diverse and representative subset of the library. This subset was utilized to create models for the T_g and θ_w, which were tested by comparing to the T_g and θ_w of the entire library. Focused libraries of polymer structures predicted to meet certain T_g and θ_w specifications were also designed. Good agreement was reported between the calculated and experimental T_g and θ_w values, even for polymers not included in the subset library. Additionally, the focused libraries were shown to be effective in identifying polymer structures within specific T_g and θ_w ranges.

Gravert et al. (*9*), used parallel synthesis to design polymeric supports for liquid-phase organic synthesis. In this work, three polymerization initiators containing α-nitrile diazene cores were utilized for block copolymer synthesis while a functionalized methacrylate initiator was used to produce graft copolymers. Five vinyl monomers were used in combination with the initiators to produce approximately 50 block and graft copolymers. Copolymer products were characterized by size exclusion chromatography, nuclear magnetic resonance and solubility in a range of solvents. Based upon this characterization, a 4-*tert*-butylstyrene-*b*-3,4-dimethoxystyrene block copolymer was selected and used successfully as a support in subsequent liquid-phase syntheses.

Takeuchi et al. (*18*) coupled combinatorial techniques with molecular imprinted polymers to develop sensors for triazine herbicides. The library consisted of a 7x7 array containing different fractions of monomers methacrylic acid (MAA) and 2-(trifluoromethyl)acrylic acid (TFMAA) with constant concentrations of the imprint molecules ametryn or atrazine. After UV-initiated polymerization the products from the sensor library were characterized by HPLC measurement of herbicide concentration. The receptor efficiency was observed to vary with monomer type: the atrazine receptor efficiency increased with MAA composition and the ametryn receptor was enhanced by increased fractions of

TFMAA. Although only monomer concentration was varied in the libraries, the authors conclude that the CM synthetic approach would be useful in analyzing other variables such as solvent, crosslinking agent, and polymerization conditions to produce optimum molecularly imprinted polymer sensors.

Dickinson *et al.* (*12*) report CM synthesis of a sensor library consisting of solvatochromic dyes dissolved in polymer. Permeation of the polymer by volatile solvents induced changes in the dye's solvation environment that were detectable by the fluorescence signal. A combination of methyl methacrylate and dimethyl-(acryloxypropyl) methylsiloxane monomers was used to create two sensor libraries. A discrete library was prepared by photo-polymerizing constant concentration solutions of the dye and monomers to produce cones of polymer at different locations on the end of a fiber-optic bundle. A second, continuous library was created by adding methyl methacrylate to the copolymer monomer as UV light was scanned across the fiber-optic bundle, producing a copolymer concentration gradient across the bundle end. Both the discrete and continuous libraries were characterized by monitoring the fluorescence response (via the fiber optic cables) as a function of exposure to saturated organic vapors. The deposition of the library directly onto the measurement probe (fiber optic) makes this work a good example of a combined CM polymer synthesis and characterization. For the particular dye and monomer used, the fluorescence response was found to be a nonlinear function of the concentration.

Newkome et al. (*26*) report a combinatorial strategy for synthesizing dendrimers with modified structure and surface chemistry. Mixtures of three branched isocyanate-based monomers, mixed over a wide range of compositions, were used to synthesize a combinatorial library of dendritic molecules. Based upon ^{13}C NMR spectra, the dendrimer products displayed varying degrees of peripheral heterogeneity, adjustable by controlling the ratios of the three isocyanate monomer groups. The methodology provides for the rapid modification of dendritic properties based upon the chemistry and distribution of peripheral surface groups. For example, some of the dendrimers were amphiphilic, displaying solubility in MeOH, H_2O, and $CHCl_3$. The degree of amphiphilicity can be adjusted to favor solubilization in one of the solvents by varying the proportion of amino versus benzyl ether surface moieties, based upon the ratio of monomer building blocks.

Combinatorial Polymer Characterization

The previous section focused on CM studies in which the production or synthesis of new polymeric materials was the primary goal. In those examples the synthesis steps were combinatorial, but subsequent characterization steps were noncombinatorial. One exception is the fluorescent sensor libraries prepared on fiber optic bundles, discussed above (*12*). In this section we describe library preparation and high-throughput screening methods for the combinatorial *characterization* of both thick (≈ 1 μm to ≈ 50 μm) and thin (< 1 μm) polymer films and coatings. Here, the primary goal is not to produce new materials, but rather to use CM to measure relevant phase behavior, wetting, and microstructural properties over a large range of parameter combinations. The variables of primary importance in characterizing the physical and chemical properties of polymers in the bulk and film state include the composition in multicomponent mixtures and composites, thickness, temperature (e.g., annealing, curing, melt processing), and substrate energy (γ_{so}).

To prepare polymer films and coatings libraries with variations in ϕ, h, T, and γ_{so}, we have found that the deposition of films with continuous gradients in each of these properties is a convenient and practical alternative to the deposition of libraries containing discrete regimes. Of course the introduction of chemical, thickness, and thermal gradients drives nonequilibrium transport processes that will eliminate the gradients over time. The timescale and lengthscale over which gradient library measurements are valid are determined in part by the magnitude of these transport fluxes. In most cases high molecular mass (27) ($M_w > 10000$ g/mol) polymers have relatively low transport coefficients, e.g., diffusivity and viscosity. Thus the mass transport and flow lengthscale and timescale are often orders of magnitude lower than those of the measurements, allowing properties to be measured near equilibrium.

Preparation of Polymer Coating and Thin Film Libraries

Thickness gradient libraries. A velocity-gradient knife coater (*21-24*), depicted in Figure 1, was developed to prepare coatings and thin films containing continuous thickness gradients. A 50 μL drop of polymer solution (mass fraction 2 % to 5 %) was placed under a knife-edge with a stainless steel blade width of 2.5 cm, positioned at a height of 300 μm and at a 5° angle with respect to the substrate. A computer-controlled motion stage (Parker Daedal) moves the substrate under the knife-edge at a constant acceleration, usually (0.5 to 1) mm/s². This causes the substrate coating velocity to gradually increase from zero to a maximum value of (5 to 10) mm/s. The increase in fluid volume passing under the knife edge with increasing substrate velocity results in films with controllable thickness gradients. Figure 2 shows *h*-gradients for polystyrene and blends of polystyrene / poly(vinylmethylether) films on Si substrates as a function of solution composition. Thin film thickness-dependent

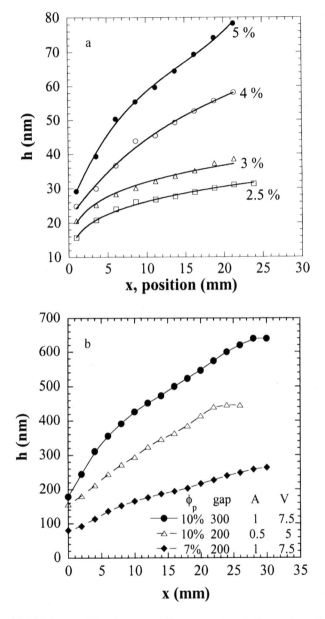

Figure 2. (a) Thickness, h (nm), versus distance, x (mm), for various h-gradient film libraries composed of polystyrene (M_w = 1800 g/mol) on Si as a function of mass fraction PS in the toluene coating solution. (b) h (nm), versus distance, x (mm), for h-gradient libraries of mass fraction 20 % PS (M_w = 96,400 g/mol)/ 80 % PVME (M_w = 119,000 g/mol) blends on Si as a function of mass fraction polymer composition in the toluene coating solution, blade substrate gap (μm), acceleration A (mm/s^2), and velocity (mm/s). Standard uncertainty in thickness is ± 3 nm. Fig 2a adapted with permission from ref. (22).

phenomena can be investigated from nanometers to micrometers employing several h-gradient films with overlapping gradient ranges. One can verify that the relatively weak thickness and temperature gradients do not induce appreciable flow in the polymer film over the experimental time scale (21,22). A unidirectional Navier-Stokes model for flow over a flat plate estimates lateral flow at a characteristic velocity of 1 μm/h at $T = 135$ °C, in response to gravitational action on the thickness gradient (28). This small flow is orders of magnitude slower than the flow induced by the physical phenomena that these libraries are designed to investigate, such as dewetting (22) and phase separation (21). To check for flow, we examined four thickness-gradient libraries before and after heating on the temperature gradient stage described below. The difference of thickness gradients across the 2 cm x 3 cm library area before and after annealing was within a standard uncertainty of ± 1.5 nm (22).

Composition gradient libraries. Three steps are involved in preparing composition gradient films: gradient mixing (Figure 3a), gradient deposition (Figure 3b), and film spreading (Figure 3c). Gradient mixing utilizes two syringe pumps (Harvard PHD2000) (29) that introduce and withdraw polymer solutions (of mass fraction $x_A = x_B = 0.05$ to 0.10) to and from a small mixing vial at rates I and W, respectively. Pump W was used to load the vial with an initial mass M_o of solution B of $M_o \approx 1$ g. The infusion and withdrawal syringe pumps were started simultaneously under vigorous stirring of the vial solution, and a third syringe, S, was used to manually extract ≈ 50 μL of solution from the vial into the syringe needle at the rate of $S = (30$ to $50)$ μL/min. At the end of the sampling process, the sample syringe contained a solution of polymers A and B with a gradient in composition, ∇x_A, along the length of the syringe needle. The relative rates of I and W were used to control the steepness of the composition gradient, e.g., dx_A/dt. The sample time, t_s, determines the endpoint composition of the gradient. The gradient produced by a particular combination of I, W, S, M_o, and t_s values was modeled by a mass balance of the transient mixing process, given elsewhere (21). This balance predicts that the composition gradient will be linear only if $I = (W + S)/2$, a prediction supported by FTIR measurements of composition. An 18 gauge needle long enough to contain the sample volume ensured that the gradient solution did not enter the syringe itself. This prevented turbulent mixing that might occur upon expansion of the solution from the needle into the larger diameter syringe.

Under the influence of the gradient in the syringe needle, ∇x_A, molecular diffusion will homogenize the composition. However, the timescale for molecular diffusion is many orders of magnitude larger than the sampling time. For example, consider gradient solutions of polystyrene (PS) ($M_w = 96.4$ kg/mol,

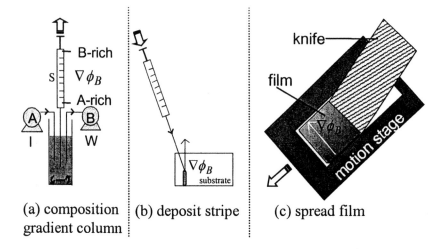

(a) composition gradient column

(b) deposit stripe

(c) spread film

Figure 3. Schematic of the composition gradient deposition process involving (a) gradient mixing, (b) deposition of stripe, and (c) film spreading. Adapted with permission from ref. (*21*).

M_w/M_n = 1.01, Tosoh Inc.) and poly(vinylmethyl ether) (PVME) (M_w = 119 kg/mol, M_w/M_n = 2.5) in toluene, a system used to characterize the ϕ-gradient deposition procedure (*21,30*). For a typical ϕ-gradient with $\Delta\phi \approx$ 0.025 mm^{-1}, ϕ_{PS} and ϕ_{PVME} change negligibly by 0.004 % and 0.001 % in the 5 min required for film deposition (*21,31*). Fluid flow in the sample syringe remains in the laminar regime, preventing turbulence and convective mixing, discussed elsewhere (*21*).

The next library preparation step (Figure 3b) is to deposit the gradient solution from the sample syringe as a thin stripe, usually (1 to 2) mm wide, on the substrate. This gradient stripe was spread as a film orthogonal to the composition gradient using the knife-edge coater described above. After a few seconds most of the solvent evaporated, leaving behind a thin film with a gradient of polymer composition. The remaining solvent was removed under vacuum during annealing, described in the next section (T-gradient annealing). Because polymer melt diffusion coefficients, *D,* are typically of order 10^{-12} cm^2/s, diffusion in the cast film can be neglected if the lengthscale resolved in measurements is significantly larger than the diffusion length, $\sqrt{Dt}$.

Composition gradient films of blends of PS/PVME and poly(D,L-lactide) (PDLA, Alkermes, Medisorb 100DL, M_w = 127,000 g/mol, M_w/M_n = 1.56) / poly(ε-caprolactone) (PCL, Aldrich, M_w = 114,000 g/mol, M_w/M_n = 1.43) were used to test the ϕ-gradient procedure. FTIR spectra were measured with a Nicolet Magna 550 and were averaged 128 times at 4 cm^{-1} resolution. The beam diameter, 500 μm (approximate), was significantly larger than the diffusion length of 3 μm (approximate) for the experimental timescale. Films (0.3 to 1) μm thick were coated on a sapphire substrate and a translation stage was used to obtain spectra at various positions on the continuous ϕ-gradient.

Figure 4a shows typical FTIR spectra for a ϕ-gradient film of PS/PVME. As position is scanned along the film, a monotonic increase in PVME absorbances, and a corresponding decrease in PS absorbances is observed. For the PS/PVME blend, compositions were measured based upon a direct calibration of the ν = 2820 cm^{-1} peak using known mixtures, yielding ϵ(2820 cm^{-1}) = 226 +/- 3 *A/hc*, where *A* = absorbance for this peak, *h* is the film thickness measured in micrometers and *c* is the molar density of PVME in moles per liter. For PDLA/PCL system, $\epsilon(\nu)$ values for pure PDLA and PCL were determined over the C-H stretch regime of (2700 to 3100) cm^{-1}, based upon $\epsilon_i(\nu)$ = $A_i(\nu)/(ch)$, where A_i is the absorbance for each peak. Unknown PDLA/PCL mass fractions were determined to within a standard uncertainty of 4 % by assuming the observed spectra were linear combinations of pure PDLA and PCL

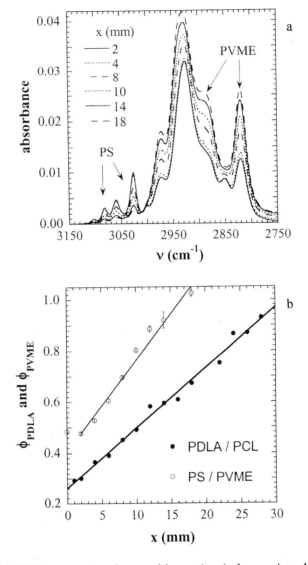

Figure 4. (a) FTIR spectra at various positions x (mm) along a φ-gradient PS/PVME library, as described in the text. PS absorptions decrease and PVME absorptions increase, monotonically, as one samples spectra across the film. (b) Mass fractions ϕ_{PVME} and ϕ_{PCL} versus position, x (mm), for typical PCL/PDLA and PS/PVME φ-gradient libraries. Composition of PS/PVME blends is calculated by calibration of the $v = 2820$ cm^{-1} PVME absorption. Composition in PDLA/PCL blends is calculated by the methodology described in the text. Coating parameters were: PS/PVME (I=0.51 mL/min, W=1.0 mL/min, S=20 μL/min, M_o=1.57 mL, sample time=94 s) and PDLA/PCL (I=0.76 mL/min, W=1.5 mL/min, S=26 μL/min, M_o=1.5 mL, sample time=95 s) Unless otherwise indicated by error bars, standard uncertainty is represented by the symbol size. Fig 4b adapted with permission from ref. (*21*).

spectra, e.g., $A_{mix} = h(\alpha\varepsilon_{PDLA}c_{PDLA} + (1-\alpha)\varepsilon_{PCL}c_{PCL})$ and α is related to the mass fraction PDLA. Figure 4b shows typical composition gradients for PDLA/PCL blends coated from CHCl$_3$ and PS/PVME blends coated from toluene. Essentially linear gradients were obtained and the endpoints and slope agree with those predicted from mass balance (*21*). It is possible to create gradient films with wider composition ranges than those shown in Figure 4, by sampling the mixing vial, Figure 3a, for longer times.

Temperature Gradient Libraries. To explore a large T range, the h- or ϕ-gradient films are annealed on a T-gradient heating stage, with the T-gradient *orthogonal* to the h- or ϕ-gradient. This custom aluminum T-gradient stage, shown in Figure 1, uses a heat source and a heat sink to produce a linear gradient ranging between adjustable end-point temperatures. End-point temperatures typically range from (160 $\pm$ 0.5) °C to (70.0 $\pm$ 0.2) °C over 40 mm, but are adjustable within the limits of the heater, cooler, and maximum heat flow through the aluminum plate. To minimize oxidation and convective heat transfer from the substrate, the stage is sealed with an o-ring, glass plate, and vacuum pump. Each two-dimensional T-h or T-ϕ parallel library contained about 1800 or 3900 state points, respectively, where a "state point" is defined by the T, h, and ϕ variation over the area of a 200X optical microscope image: $\Delta T = 0.5$ °C, $\Delta h = 3$ nm, and $\Delta\phi = 0.02$. These libraries allow T, h, and ϕ-dependent phenomena, e.g., dewetting, order-disorder, and phase transitions, to be observed *in situ* or post-annealing with relevant microscopic and spectroscopic tools.

Surface energy gradients. In many polymer coating and thin film systems, there is considerable interest in studying the film stability, dewetting, and phase behavior on substrates with surface energies varying between hydrophilic and hydrophobic extremes. Therefore, a gradient-etching procedure has been developed in order to produce substrate libraries with surface energy, γ_{so}, continuously varied from hydrophilic to hydrophobic values (33). The gradient-etching procedure involves immersion of a passivated Si-H/Si substrate (Polishing Corporation of America) into a 80 °C Piranha solution (*34*) at a *constant immersion rate*. The Piranha bath etches the Si-H surface and grows an oxide layer, SiO$_x$/SiOH, at a rate dependent on T and the volume fraction H$_2$SO$_4$ (*34*). A gradient in the conversion to hydrophilic SiO$_x$/SiOH results because one end of the wafer is exposed longer to the Piranha solution. After immersion, the wafer is withdrawn rapidly ($\approx$10 mm/s), rinsed with deionized water, and blown dry with N$_2$. Typical deionized water contact angles are shown in Figure 5. By preparing several gradient substrates covering overlapping ranges of hydrophilicity, it is possible to screen a large range in surface energy, from hydrophilic ($\theta_w \approx 0°$) to hydrophobic ($\theta_w \approx 90°$) values of the water contact angle.

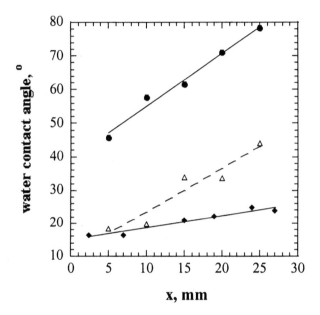

Figure 5. Deionized water contact angle versus position (mm) for gradient-etched SiH/Si substrates as a function of both immersion rate and Piranha solution H_2SO_4 composition. Immersion rate and mass fraction H_2SO_4 are as follows: circles: 2.0 mm/s and 30 %; triangles: 0.1 mm/s and 30 %; diamonds: 2.0 mm/s and 40 %. Standard uncertainty in contact angle is ± 2°.

In another procedure for varying substrate energy, developed by other authors, mixed self-assembled monolayers (SAMs) of alkanethiolates are deposited with a composition gradient (35). In this procedure, ω-substituted alkanethiolates with different terminal groups, e.g., -CH$_3$ versus -COOH, cross-diffuse from opposite ends of a polysaccharide matrix deposited on top of a gold substrate. Diffusion provides for the formation of a SAM with a concentration gradient between the two thiolate species from one end of the substrate to another, resulting in controllable substrate energy gradients. The polysaccharide matrix is removed after a period of time, halting the diffusion process. These gradient SAM substrates were subsequently used to investigate the effect of surface energy on phase separation of immiscible polymer blends (36).

Fundamental Property Measurement with Combinatorial Polymer Coating and Film Libraries

Thin Film Dewetting. Figure 6 shows a composite of optical microscope images of a *T-h* library of PS (Goodyear, M_w = 1900 g/mol, M_w/M_n = 1.19) on a SiO$_x$/SiOH substrate (22). The thickness ranges from (33 to 90) nm according to $h = 33.1x^{0.30}$ (1 < x < 28) mm and 85 °C < T < 135 °C. The images, taken 2 h after initiation of dewetting, show wetted and dewetted regimes that are visible as dark and bright regions, respectively, to the unaided eye. Repeated examination of combinatorial *T-h* libraries at thicknesses ranging from (16 to 90) nm indicates three distinct thickness regimes with different hole nucleation mechanisms. For h > (55 ± 4) nm, discrete circular holes in the film nucleate via heterogeneities (e.g., dust) and grow at a rate dependent on T (quantification of the rate is given in Figure 7).

Below h ≤ (55 ± 4) nm, there is a sharp and temperature independent transition to a regime where irregular, asymmetrical holes nucleate and grow more slowly than at h > 55 nm. In the regime (33 < h < 55) nm, the heterogeneous and capillary instability nucleation mechanisms compete. The asymmetrical holes present in this regime are surrounded by bicontinuous undulations in the film surface, with a characteristic spacing of 7 µm, as indicated by optical microscopy (22). AFM indicates a roughened surface with correlated surface undulations. A similar structure consisting of asymmetrical holes that break up into a bicontinuous pattern at late stage, termed an "intermediate morphology", has been observed recently for 12 nm thick films of poly(styrene-*ran*-acrylonitrile) (37). Below h ≈ 33 nm, another transition in structure and nucleation is apparent. Here, holes are nucleated by capillary instability and grow more quickly than in the region 33 nm < h < 55 nm.

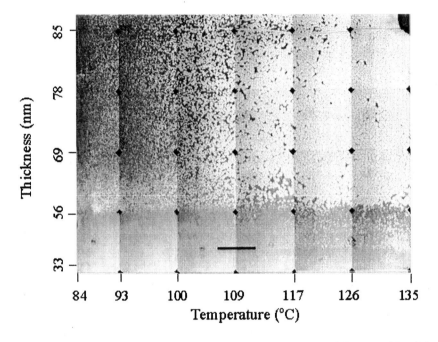

Figure 6. Composite of optical images of a *T-h* combinatorial library of PS (*M*$_w$ = 1800 g/mol) on silicon, *t* = 2 h, 25x magnification. Scale bar = (2.0 ± 0.1) mm. The thickness scale is a power law function (given in text), reflecting the nonlinear thickness gradient deposition procedure. Adapted with permission from ref. (*22*).

Systematic studies of the temperature dependence of dewetting rates have not been reported to our knowledge. However, through its effect on viscosity, T is expected to have a strong influence on hole drainage rates. By using automated optical microscopy (Figure 1), a 5 x 5 grid of images covering the T and h range of the library was collected every 5 min for 2 h. Automated image analysis (22) of hole area as a function of T, h, and t assays a broad range of dewetting rates in a single experiment, shown in Figure 7. The inset to Figure 7 shows the raw dewetted area fraction, A_d, versus t at various temperatures from the T-h library in Figure 6 for $h = 79$ nm. The entire set of A_d vs. t profiles for h > 55 nm can be fitted with $A_d = A_{d\infty} + (A_{do} - A_{d\infty})\exp(-(t-t_o)/\tau)$, where A_{do} and $A_{d\infty}$ are the dewet fractions at $t = t_o$ and $t = \infty$, τ is the dewetting time constant, and t_o is a time delay for nucleation. As shown in Figure 7, in reduced units of $(A_{d\infty} - A_d)/(A_{d\infty} - A_{do})$ versus $(t - t_o)/\tau$, the hole drainage profiles collapse onto the universal exponential curve given above. Figure 7 contains T and h data over a large range (92 < T < 135) °C and (59 < h < 86) nm, and τ ranges from 2100 s (high temperatures) to 113,000 s (low temperatures). This T, h superposition for dewetting rates, previously unreported, reflects variations in the film viscosity with T and h, and could be missed altogether by relying solely upon limited numbers of samples.

Phase Behavior. Figure 8 presents a photograph of a typical temperature-composition library of the PS/PVME blend (discussed in the composition gradient preparation section above) after 90 min of annealing. As Figure 8 indicates, the lower critical solution temperature (LCST) cloud point curve can be seen with the unaided eye as a diffuse boundary separating one-phase and two-phase regions. Cloud points measured on bulk samples with conventional light scattering are shown as discrete data points and agree well with the cloud point curve observed on the library (21,38). The diffuse nature of the cloud point curve reflects the natural dependence of the microstructure evolution rate on temperature and composition. Near the LCST boundary the microstructure size gradually approaches optical resolution limits (1 μm), giving the curve its diffuse appearance. Based upon a bulk diffusion coefficient of $D \approx 10^{-17}$ m²/s, the diffusion length ($\sqrt{Dt}$) for a 2 h anneal is 270 nm. In Figure 8 each pixel covers about 30 μm, which is over 100 times the diffusion length, and ϕ-gradient-induced diffusion has a negligible effect on the observed LCST cloud point curve. The combinatorial technique employing T-ϕ polymer blend libraries allows for rapid and efficient characterization of polymer blend phase behavior (cloud points) in orders of magnitude less time than with conventional light scattering techniques.

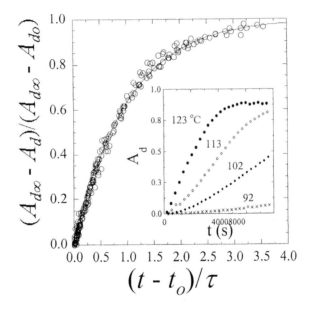

Figure 7. *T-h-t* superposition of dewet area fraction data onto a universal curve, (92 < *T* < 135) °C, (59 < *h* < 86) nm. Inset: raw dewet area fraction vs. time, *h* = 79 nm. Adapted with permission from ref. (*22*).

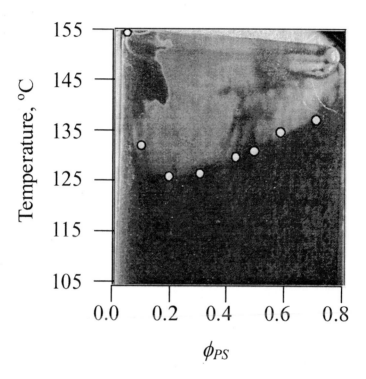

Figure 8. Digital optical photographs of a PS/PVME T-ϕ library after 91 min of annealing, showing the LCST cloud point curve visible to the unaided eye. The library wafer dimension is 31 mm x 35 mm and the film thickness varies from approximately (400 to 600) nm from low to high ϕ_{PS} values. White circles light-scattering cloud points measured on separate uniform samples.

Block copolymer segregation and surface morphology. The morphology of symmetric diblock copolymer thin films has been studied extensively with traditional techniques (*39-52*). These studies found surface induced formation of lamellae with thickness equal to the equilibrium lamellar thickness L_o parallel to the substrate. The lamellae form smooth films when the total film thickness h is equal to an integral multiple m of L_o, $h \approx mL_o$ for one block segregating to both the substrate and air interfaces and $h \approx (m + \frac{1}{2})L_o$ for one block preferring the substrate and the other preferring the air interface. When h deviates from these values, holes or islands of height L_o are found to form on the film surface in order to reduce the system energy. Although there has been significant previous investigation of these systems, almost no research on the factors controlling the lateral dimensions of these surface features has been reported. In addition, no investigation of the transition regions between the hole and island formations has been performed due to the difficulty in accurately controlling film thickness with traditional spin-coating techniques. These types of questions are ideal for combinatorial methods where it is possible to produce continuous h-gradient films of symmetric diblock copolymers with various molecular masses (*23,24*).

Gradients in h of symmetric polystyrene-*b*-poly(methyl methacrylate) (PS-*b*-PMMA) with three different molecular masses were produced using the knife-edge flow coating technique described above. After characterization of h with UV-visible interferometry, the films were annealed at 170 °C for up to 30 h to allow lamellar organization. The resulting morphologies were characterized with optical microscopy (OM) and AFM. Plate 1a presents a true color optical micrograph of a M_w = 26,000 g/mol gradient film showing morphological changes associated with the addition of two lamellae to the surface of the film. Labels denote approximate h values corresponding to $h \approx (m + \frac{1}{2})L_o$ for $m = 4, 5$ and 6. The morphology evolves from a smooth film to circular islands to a bicontinuous hole/island region to circular holes back to a smooth film from left to right in the micrograph and repeats twice. This morphology was observed to consistently repeat for h ranging from $2.0L_o$ to $6.5L_o$ for all molecular masses examined. Notably, the smooth regions of the film form a significant fraction of the morphology, corresponding to a thickness range Δh deviating significantly from an integral multiple of L_o. This Δh is confirmed in the optical micrograph presented in Plate 1b where the smooth region of a M_w = 104,000 g/mol PS-*b*-PMMA gradient film is displayed. The orange features on the left are holes and the yellow-green features on the right are islands and the smooth region changes color from purple to blue-green indicating an h increase. The value of Δh is $\approx 0.28L_o$ and invariant within standard uncertainty for all M_w and h investigated. This effect is interpreted to arise from a brush-like stretching and compression of block copolymer chains in the outer lamella as the chain density varies with h. Therefore islands and holes form only when the free energy penalty of chain deformation becomes so large that the defect structures are more energetically favorable.

The lateral size of the surface patterns, which can be remarkably large when compared to L_o or h, was also investigated using the h-gradient block copolymer film libraries. Figure 9 shows AFM micrographs of the surface of PS-b-PMMA gradient films with M_w of 26,000 g/mol (Figure 9a), 51,000 g/mol (Figure 9b), and 104,000 g/mol (Figure 9c), annealed for 30 h at 170 °C. These micrographs demonstrate that the lateral scale of the surface features decreases with increasing M_w. This fact is quantified by obtaining 2D Fourier transforms of the micrographs and extracting a characteristic peak wavevector (q^*) for each M_w. Figure 9d contains a plot of $\lambda \equiv (1/q^*)$ vs. L_o for samples annealed for both 6 h and 30 h and the lines correspond to power law fits yielding the relation λ (μm) $\sim L_o^{-2.5}$ or correspondingly λ (μm) $\sim M_w^{-1.5}$. This decrease of λ with increasing M_w suggests that as M_w of the outer block copolymer layer increases, its viscoelastic nature increases the free energy cost of forming large-scale surface patterns. The large-scale pattern formation in block copolymer films is therefore tentatively ascribed to the increased surface energy required to deform the surface of the block copolymer layers with increasing M_w.

Organic Light-emitting diodes. Finally, we mention here two characterization studies of the optimization of organic light emitting diodes (OLEDs). Schmitz and coworkers (*14-16*) used a masked deposition technique to produce thickness gradients in both the organic hole transport layers and the inorganic electron transport and emitting layer. OLEDs with single-gradient and orthogonal two-dimensional gradient structures were produced in order to evaluate the effects of the various layer thicknesses on the device efficiency. An optimal thickness for both the hole and electron transporting layers was reported. Likewise Gross *et al.* (*17*) have reported the use of combinatorial methods to investigate the performance of doped (oxidized) π-conjugated polymers in OLEDs. In these devices the polymers serve as hole transport layers but an energy barrier for hole injection exists between the polymeric material and the inorganic anode. By varying the oxidation level of the polymer this energy barrier can be reduced to lower the device working voltage. The effect of oxidation was studied by electrochemically treating the polymer to create a continuous gradient in the oxidation level of the polymer. A gradient in thickness was created orthogonal to the gradient in oxidation to explore variations of both properties simultaneously. For this reason, this study represents a cross between both combinatorial synthesis (oxidation steps) and process characterization (thickness gradient deposition). The gradient libraries were characterized by monitoring the efficiency and onset voltage of OLEDs fabricated on the gradients.

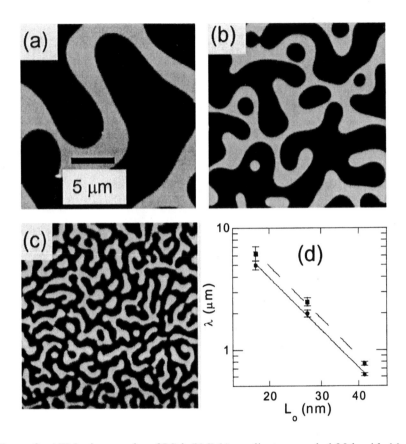

Figure 9. AFM micrographs of PS-b-PMMA gradients annealed 30 h with M_w of (a) 26,000 g/mol, (b) 51,000 g/mol and (c) 104,000 g/mol showing the decrease in surface feature size with increasing M_w. (brighter regions correspond to higher topography, scale bar applies to all micrographs). d) Plot of λ vs. L_o for samples annealed for 6 h (circle, solid line) and 30 h (square, dashed line) showing the power law dependence of λ on L_o and M_w.

Conclusions

In this brief review we have presented recent advances in which combinatorial methodologies have been used to efficiently measure chemical and physical properties of polymers over large regimes of variable space. Methodologies for both synthesis and characterization allow for both materials discovery and discovery of new models and structure-processing-property relationships. Polymer synthesis with combinatorial methods relies primarily on reactant mixing and diversity to create discrete libraries for investigating reactant structure and composition variations. In the future, techniques will be developed to allow library variation in processing variables like temperature, UV exposure, and pressure during synthesis. Several recent developments were presented in the combinatorial characterization of polymers using high-throughput libraries of films and coatings. As examples, we have developed four novel techniques for preparing continuous gradient polymer libraries with controlled variations in temperature, composition, thickness, and substrate surface energy. The use of these new library techniques facilitates characterization of polymer blend phase behavior, thin-film dewetting, and block copolymer order-disor der transitions.

References

1. Reddington, E.; Sapienza, A.; Gurau, B.; Viswanathan, R.; Sarangapani, S.; Smotkin, E.; Mallouk, T. *Science* **1998**, *280*, 1735.
2. Xiang, X.-D.; Sun, X.; Briceno, G.; Lou, Y.; Wang, K.-A.; Chang, H.; Wallace-Freedman, W. G.; Chen, S.-W.; Schultz, P. G. *Science* **1995**, *268*, 1738.
3. Wang, J.; Yoo, Y.; Gao, C.; Takeuchi, I.; Sun, X.; Chang, H.; Xiang, X.-D.; Schultz, P. G. *Science* **1998**, *279*, 1712.
4. Sun, X.-D.; X.-D., X. *Appl. Phys. Lett.* **1998**, *72*, 525.
5. Danielson, E.; Golden, J. H.; McFarland, E. W.; Reaves, C. M.; Weinberg, W. H.; Wu, X. D. *Nature* **1997**, *389*, 944-48.
6. Danielson, E.; Devenney, M.; Giaquinta, D. M.; Golden, J. H.; Haushalter, R. C.; McFarland, E. W.; Poojary, D. M.; Reaves, C. M.; Wenberg, W. H.; Wu, X. D. *Science* **1998**, *279*, 837-39.
7. Jandeleit, B.; Schaefer, D. J.; Powers, T. S.; Turner, H. W.; Weinberg, W. H. *Angew. Chem. Int. Ed.* **1999**, *38*, 2494.
8. Klein, J.; Lehmann, C. W.; Schmidt, H.-W.; Maier, W. F. *Angew. Chem. Int. Ed.* **1998**, *37*, 3369.
9. Gravert, D. J.; Datta, A.; Wentworth, P.; Janda, K. D. *J. Am. Chem. Soc.* **1998**, *120*, 9481.

10. Brocchini, S.; James, K.; Tangpasuthadol, V.; Kohn, J. *J. Am. Chem. Soc.* **1997**, *119*, 4553.
11. Brocchini, S.; James, K.; Tangpasuthadol, V.; Kohn, J. *J. Biomed. Mater. Res.* **1998**, *42*, 66.
12. Dickinson, T. A.; Walt, D. R.; White, J.; Kauer, J. S. *Anal. Chem.* **1997**, *69*, 3413.
13. Reynolds, C. H. *J. Comb. Chem.* **1999**, *1*, 297.
14. Schmitz, C.; Posch, P.; Thelakkat, M.; Schmidt, H. W. *Phys. Chem. Chem. Phys.* **1999**, *1*, 1777.
15. Schmitz, C.; Thelakkat, M.; Schmidt, H. W. *Adv. Mater.* **1999**, *11*, 821.
16. Schmitz, C.; Posch, P.; Thelakkat, M.; Schmidt, H. W. *Macromol. Symp.* **2000**, *154*, 209.
17. Gross, M.; Muller, D. C.; Nothofer, H. G.; Sherf, U.; Neher, D.; Brauchle, C.; Meerholz, K. *Nature* **2000**, *405*, 661.
18. Takeuchi, T.; Fukuma, D.; Matsui, J. *Anal. Chem.* **1999**, *71*, 285.
19. Terrett, N. K. *Combinatorial Chemistry*; Oxford: Oxford, 1998.
20. Kennedy, K.; Stefansky, T.; Davy, G.; Zackay, V. F.; Parker, E. R. *J. Appl. Phys.* **1965**, *36*, 3808-3810.
21. Meredith, J. C.; Karim, A.; Amis, E. J. *Macromolecules* **2000**, *33*, 5760-5762.
22. Meredith, J. C.; Smith, A. P.; Karim, A.; Amis, E. J. *Macromolecules* **2000**, *33*, 9747-9756.
23. Smith, A. P.; Meredith, J. C.; Douglas, J. F.; Amis, E. J.; Karim, A. *Phys. Rev. Lett.* **2000**, in press.
24. Smith, A. P.; Douglas, J. F.; Meredith, J. C.; Amis, E. J.; Karim A. *J. Polym. Sci. B: Polym. Phys.* **2000**, in press.
25. Hanak, J. J. *J. Mat. Sci.* **1970**, *5*, 964.
26. Newkome, G. R.; Weis, C. D.; Moorefield, C. N.; Baker, G. R.; Childs, B. J.; Epperson, J. *Angew. Chem. Int. Ed.* **1998**, *37*, 307-310.
27. According to ISO 31-8, the term "molecular weight" has been replaced by "relative molecular mass", symbol Mr. The conventional notation, rather than the ISO notation, has been employed for this publication.
28. Leal, L. G. *Laminar Flow and Convective Transport Processes*; Butterworth-Heinemann: Boston, 1992.
29. Certain equipment and instruments or materials are identified in the paper in order to adequately specify the experimental details. Such identification does not imply recommendation by the National Institute of Standards and Technology, nor does it imply the materials are necessarily the best available for the purpose.
30. Daivis, P. J.; Pinder, D. N.; Callaghan, P. T. *Macromolecules* **1992**, *25*, 170-178.

31. The diffusive flow rate of PS and PVME were calculated as $J = L\pi r^2 D_i \rho (d\phi_i/dx)_{max}$, where ρ is the solution density, $r=2.3$ mm is the syringe diameter, and $L=4.2$ mm is the length of the fluid column in the syringe. We estimate $\Delta\phi_i$ as $(Jt)/(x_p L\pi r^2 \rho)$, where $x_p = 0.08$ is the total polymer mass fraction in solution.

32. Budtov, V. P. *Polym. Sci. USSR* **1967**, *9*, 854-862.

33. Ashley, K.; Meredith, J. C.; Karim, A.; Raghavan, D. *Langmuir* **2000**, in preparation.

34. *Handbook of Semiconductor Wafer Cleaning Technology*; Kern, W., Ed.; Noyes Publications: Park Ridge, NJ, 1993.

35. Liedberg, B.; Tengvall, P. *Langmuir* **1995**, *11*, 3821-3827.

36. Genzer, J.; Kramer, E. J. *Europhys. Lett.* **1998**, *44*, 180-185.

37. Masson, J. L.; Green, P. F. *J. Chem. Phys.* **1999**, *112*, 349.

38. Meredith, J. C.; Karim, A.; Amis, E. J. *Adv. Mater.* **2000**, *submitted*.

39. Hasegawa, H.; Hashimoto, T. *Macromolecules* **1985**, *8*, 589.

40. Henkee, C. S.; Thomas, E. L.; Fetters, L. J. *J. Mater. Sci.* **1988**, *23*, 1685.

41. Russell, T. P.; Coulon, G.; Deline, V. R.; Miller, D. C. *Macromolecules* **1989**, *22*, 4600.

42. Green, P. F.; Christensen, T. M.; Russell, T. P.; Jerome, R. *Macromolecules* **1989**, *22*, 2189.

43. Anastasiadis, S. H.; Russell, T. P.; Satija, S. K.; .Majkrzak, C. F. *J. Chem. Phys.* **1990**, *92*, 5677.

44. Ausserre, D.; Chatenay, D.; Coulon, G.; Collin, B. *J. Phys. France* **1990**, *51*, 2571.

45. Coulon, G.; Ausserre, D.; Russell, T. P. *J. Phys. France* **1990**, *51*, 777.

46. Russell, T. P.; Menelle, A.; Anastasiadis, S. H.; Satija, S. K.; Majkrzak, C. F. *Macromolecules* **1991**, *24*, 6263.

47. Green, P. F.; Christensen, T. M.; Russell, T. P. *Macromolecules* **1991**, *24*, 252.

48. Russell, T. P.; Menelle, A.; Anastasiadis, S. H.; Satija, S. K.; Majkrzak, C. F. *Macromol. Chem., Macromol. Symp.* **1992**, *62*, 157.

49. Menelle, A.; Russell, T. P.; Anastasiadis, S. H.; Satija, S. K.; Majkrzak, C. F. *Phys. Rev. Lett.* **1992**, *68*, 67.

50. Cai, Z.; Huang, K.; Montano, P. A.; Russell, T. P.; Bai, J. M.; Zajac, G. W. *J. Chem. Phys.* **1993**, *93*, 2376.

51. Coulon, G.; Daillant, J.; Collin, B.; Benattar, J. J.; Gallot, Y. *Macromolecules* **1993**, *26*, 1582.

52. Mayes, A. M.; Russell, T. P.; Bassereau, P.; Baker, S. M.; Smith, G. S. *Macromolecules* **1994**, *27*, 749.

Chapter 3

The Compositional Spread Approach to High-Dielectric Constant Materials and Materials for Integrated Optics

L. F. Schneemeyer[*], R. B. van Dover[*], C. K. Madsen, and C. L. Claypool

Bell Laboratories, Lucent Technologies, 600 Mountain Avenue, Murray Hill, NJ 07974-0636

Introduction

High throughput synthetic approaches combined with rapid screening techniques to evaluate materials performance – so called combinatorial approaches – offer important advantages for materials investigations. Optical communications and the on-going electronic revolution are driving interest in new inorganic solid state materials (1). With ever shortening product life cycles, effective research methodologies are also of great interest in industrial research. Efficient strategies for materials investigation can greatly improve the odds of a successful materials investigation., particularly when the sample materials are prepared using methods similar to those that would be employed in the actual application.

Combinatorial approaches to the synthesis of large collections, or "libraries", of molecules with possible biological activity have become a standard tool in the drug discovery and optimization process. Together with rapid evaluation of biological activity using various well-developed automated screening approaches, combinatorial chemistry has had significant impact on searches for molecular entities that match particular profiles of desired structure-property relationships.

Because similar incentives exist compelling researchers to identify inorganic materials for specific applications with greater speed, analogous high-throughput synthetic techniques together with efficient screening approaches are of increasing importance in solid state materials investigations. The traditional approach to new materials investigations is typically extremely time consuming and labor intensive. A materials chemist requires about a day to synthesize a single sample and characterization of that sample may take many more days. Clearly, increasing the number of materials that are studied should improve the odds of a breakthrough materials discovery while certainly leading to an increased understanding of composition/property relationships.

Background

The concept of the preparation of large numbers of samples simultaneously has been around since at least the late 1960's. Sawatzky and Kay at IBM used cosputtering to prepare gadolinium iron garnet films in which defect concentrations varied along the length of the film to correlate with structural and magnetic properties (2). This efficient synthetic approach eliminated the time and expense of preparing multiple targets which would have been needed had samples in the series been grown one at a time. In addition, any ambiguities which might have arisen from run-to-run variations were eliminated. However, sample evaluation was still carried out manually although the optical characterization relevant to this study was not particularly time consuming.

A broader view to the potential power of a parallel synthesis scheme was proposed by Hanak of RCA Labs in 1970 (3). Hanak carried out reactive sputtering from a sectored target as illustrated in Figure 1. A broad range of experimental systems, including superconductors and magnetic materials were studied by Hanak using this approach (4, 5). However, computer control and automated measurements were not yet available at that time, largely because computers were still very expensive and not widely available.

The importance of combinatorial chemistry in the pharmaceutical industry motivated Xiang and Schultz to revisit high-throughput synthesis and screening approaches as applied to materials investigations (6). Studies of high temperature cuprate superconductors, colossal magnetoresistance (CMR) manganites and investigations of new phospors demonstrated the power of the approach. In this discrete combinatorial synthesis, DCS, approach, arrays of samples were prepared by multilayer thin film depositions through masks followed by a high temperature annealing step to achieve mixing and produce homogeneous samples. This high temperature anneal often results in the crystallization of samples. Automated screening allows rapid evaluation of materials properties.

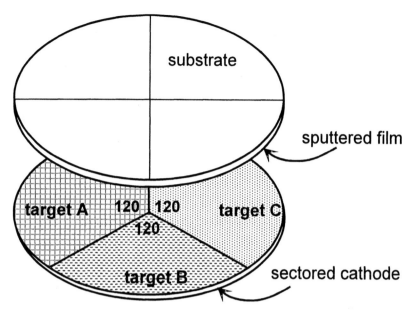

Figure 1. Sputter target layout used for ternary phase spread deposition in the Hanak approach.

While a combinatorial-type approach to a materials study can quickly produce large numbers of samples and a large quantity of data, successful application of such an approach requires the thoughtful application of traditional scientific principles as outlined in Figure 2. A well-defined problem with an explicit goal, a feasible and appropriate synthetic strategy and a measurement and evaluation protocol are elements of a proper materials investigation. Understanding the issues inherent in a particular problem is vital. Samples should preferably be prepared in a form that is meaningfully related to the form in which they will ultimately be used. Samples should be screened by measurements related to the property of interest. Ideally, the results of measurements can be reduced to a scalar figure of merit (FOM) to allow the visual discerning of trends. Also, the time required to evaluate samples should be comparable to the time required by the synthesis step.

Even though large numbers of samples can be studied in a combinatorial-type approach, the large number of possible combinations of elements can quickly overwhelm an investigation even before various possible processing parameters are considered. Thus, the extent of a study must be rationally constrained. For example, there are 720 different combinations of 10 elements taken 3 at a time. Constraints must be applied to confine a study within manageable limits.

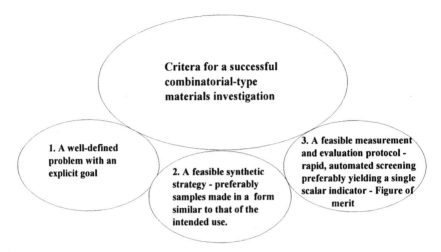

Figure 2. Criteria for the successful use of high-throughput synthesis/screening.

The Continuous Compositional Spread (CCS) Approach

While the DCS approach produces samples in thin-film form, the high temperature annealing step eliminates this approach for studies of low temperature metastable or amorphous materials. We have developed the continuous compositional spread (CCS) approach (7) for investigations of materials deposited at relatively low temperatures. In this section, the CCS approach is described and its use in the discovery of a new thin film high dielectric constant material with high breakdown fields and low leakage currents is discussed.

In the CCS approach, we use off-axis cosputtering to produce binary or ternary composition spreads as illustrated in Figure 3. In off-axis deposition onto a fixed substrate, the thickness of the deposited film decreases approximately exponentially with distance from the gun. Test runs of Si-O and Ta-O deposition demonstrated that high quality oxide thin films could be obtained by reactive sputtering in an off-axis geometry using 2 inch metallic targets in planar magnetron sputter guns (U.S. Gun II, US, Inc., Campbell, CA). A crucial factor, however, was the application of an rf field to the growing film during deposition. This bias causes ion bombardment of the growing film, probably

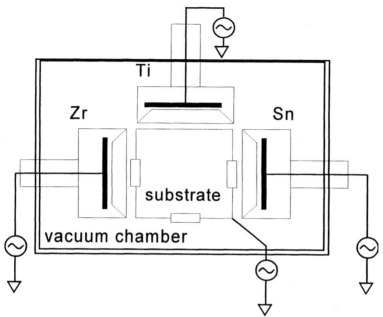

Figure 3. Schematic drawing of the off-axis deposition system used for the CCS approach.

enhancing the surface mobilities and producing a denser film. Our system allows us to codeposit up to three metals so that we can deposit a large portion of a pseudoternary oxide phase diagram in a single run. Relatively high oxygen partial pressures, 10-40%, were used to ensure fully oxidized films. Note that these phase spreads have inherent thickness variations of approximately a factor of two, a limitation of this method.

To know the composition of the film at any position on the phase spread, the off-axis deposition of SiO_x was studied and modeled. As noted earlier, thickness decreases exponentially with distance from the gun. However, the 2 inch planar magnetron sources used for these experiments are not point sources making the model somewhat more complicated. Still, fits good to 2-5% were obtained. Overall sputtering rates as well as the decay length varies with sputtering parameters including power, partial pressure of oxygen and total pressure (typically 10 to 50 mtorr). For each material to be sputtered, thickness calibrations measured using profilimetry were carried out for several sets of conditions. Compositions obtained using these models were verified by Rutherford Backscattering (RBS) measurements at several points on the sample.

Table I. Comparison of high dielectric constant materials.

Material	ε_r	E_{br} MV /cm	FOM $\mu C/cm^2$	J_{leak} A/cm^2	Comments
SiO$_2$	4.0	10	3.5 (typical)	6×10^{-12} (typical)	standard material. ε_R is low
TaO$_x$	23	4	8.1 (typical)	6×10^{-10} (best)	moderate ε, good for 1 generation
(Ba,Sr)TiO$_3$ BST	200	1	18 (typical)	1×10^{-8} (typical)	etch? electrodes? requires $T_{deposition} > 700\ °C$
Zr-Sn-Ti-O a-ZTT	62	4.4	24.3 (typical)	2×10^{-9} (best ever)	Van Dover and Schneemeyer, IEEE Electron Device Letters **19**, 329 (1998).
Hf-Sn-Ti-O	58	4	19 (conservative estimate)	1×10^{-6} (not optimized)	This work Hf$_{.20}$Sn$_{.05}$Ti$_{.75}$O$_2$

A variety of substrates, those suitable for any sputter deposition technique, can be used for the CCS approach. We find silicon wafers, <100> Si, which are very flat, clean and relatively inexpensive to be particularly convenient. Blanket coats of buffer layers, metals such as platinum, aluminum or Ti/TiN, etc., can be coated on the wafer prior to the deposition of the phase spread.

The value of the CCS approach was demonstrated by our discovery of amorphous, high dielectric constant, high breakdown field, low leakage materials in the Zr-Ti-Sn-O pseudoternary phase diagram. As integrated circuits continue to migrate to higher levels of integration, it appears likely that alternative high dielectric constant materials will be needed to replace silicon oxide, a-SiO_x. At present, designers are resorting either to extremely thin or highly nonplanar a-SiO_x films in capacitor structures used for dynamic random access memory, DRAM. A useful figure of merit corresponding physically to the maximum charge per unit area that can be stored on a capacitor made of a given material can be defined as $\varepsilon_1 \varepsilon_0 E_{BR} = FOM$, where ε_0 is the permittivity of free space and E_{BR} is the breakdown field (8). Table I contains the relevant data for materials of current interest as high dielectric constant replacement for a-SiO_x. While each of these candidate materials has certain advantages, each also suffers limitations. For example, a-TaO_x, amorphous tantalum oxide films, can be deposited at low temperatures, < 450° C, and can have high breakdown fields and low leakage currents (9). However, its dielectric constant is only modestly higher that that of a-SiO_x, making it a short-term improvement at best.

We confined our CCS search for high dielectric constant, high FOM thin film materials to oxides deposited at low temperatures, < 400° C, consistent with the constraints of backend processing and therefore amorphous. We also limited ourselves to elements compatible with existing IC fabrication processes favoring those already accepted in Si fabrication facilities. Those favored elements include Si, Al, O, N, Ti and W.

About 30 different elemental composition spreads were evaluated in our initial survey. Of these, the Zr-Ti-Sn-O system showed evidence for an extensive high FOM region as well as reasonably high specific capacitance values and low leakage, and thus was selected for more detailed examination. Figure 4 shows

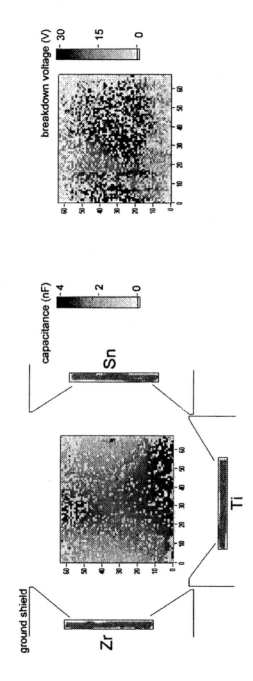

Figure 4. Raw capacitance and breakdown voltage data for a-Zr-Ti-Sn-O as measured.

Figure of Merit

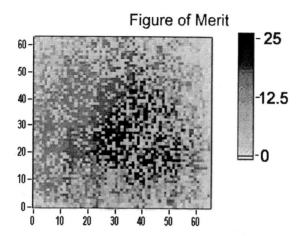

Figure 5. Raw figure of merit data for the system Zr-Ti-Sn-O deposited with guns in the same configuration as Figure 4.

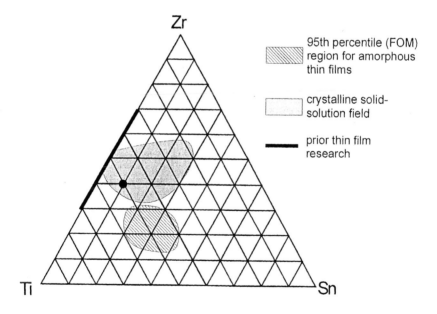

Figure 6. The FOM data for the system Zr-Ti-Sn-O plotted on a conventional ternary phase diagram. Also shown is the single phase region for the crystalline ceramic of composition $Zr_{1-x}Sn_xTiO_4$.

the raw capacitance and breakdown voltage as a function of position on the substrate. The positions of the sputter guns are shown schematically. While our films are not of uniform thickness, thickness varies gradually with position and thus valid qualitative trends can be discerned.

A thickness independent FOM can be determined by a point-by-point evaluation of the product CV_{BR}/A, the capacitance times the breakdown voltage divided by the area of the capacitor structure measured. This quantity is conveniently equal to $\varepsilon_1\varepsilon_0 E_{BR}$, the FOM discussed above. The FOM data for the Zr-Ti-Sn-O systems in the same orientation as Figure 4 is shown in Figure 5 together with the leakage current values measured at a stored charge density of 7 $\mu C/cm^2$. A distinct region of high FOM and a separate region of low leakage are clearly seen. Figure 6 shows the high FOM region (95th percentile contour) mapped onto a conventional ternary phase diagram. This figure also shows the single phase region for the low-loss dielectric ceramic material, $Zr_{1-x}Sn_xTiO_4$, which is used for filter elements in wireless circuits (10). Film studies involving compositions inspired by only by the single phase ceramic region (11, 12) miss both the high FOM region and the low loss regions of the phase diagram. Because our high throughput search approach allowed us to search broadly, we found excellent properties in an unexpected region of the composition space. The composition a-$Zr_{.2}Ti_{.6}Sn_{.2}O_x$ was identified as a high specific capacitance, high breakdown field and low leakage thin film material that represents a promising solution to the problem of dielectric for future generations of embedded DRAM.

The Optics Explosion

Demand for Internet access and other broadband services has exploded in the late 1990's leading to a push for higher and higher lightwave transmission capacity. The two approaches to providing that capacity are high transmission rates and also wavelength division multiplexing (WDM). The growth rates in optical communication, which are >50% annually, are creating demands for new components for optical communications (13).

A dense wavelength division multiplexing, DWDM, optical fiber transmission system increases transmission capacity by combining multiple wavelengths of light onto a single optical fiber which are then separated at the other end as illustrated schematically in Figure 7. Multiplexing and demultiplexing, mux and demux, is accomplished using a PLC devices called arrayed waveguide grating routers, AWG's. For long-haul systems, either an erbium or Raman optical

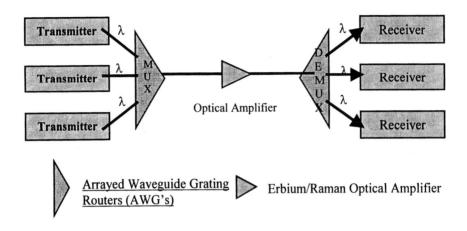

Figure 7. Simplified schematic drawing of a WDM system.

amplifier is used to boost the signal intensity during transmission. Each wavelength provides a separate channel, which can carry voice, data or video information.

While long distance, so-called long haul, transmission systems were the original focus for WDM, an evolution is envisioned in which systems will move from WDM transport to WDM networks. This evolution will involve first the use of fixed, and then, in the more distant future, reconfigurable add-drop nodes. Also, there will be more wavelengths, higher bit rates and more powerful amplifiers as well as new optical amplifiers operating over various regions of the Allwave™ fiber spectrum extending from about 1200 to 1700 nm.

Integrated optical components fabricated using variants of well-established silicon processing technology, known as silicon optical bench technology, are likely to play significant roles in future system developments. Of course, several technologies are competing to provide critical capabilities in areas that include routers, reconfigurable add/drop multiplexers, dynamic gain equalizers, dispersion compensators and optical amplifiers (13). Bulk-type devices such as MEM's devices and thin-film filters, fiber-type devices such as fiber couplers and planar-type devices such as silicon optical bench and semiconductor devices can all provide various aspects of the needed functionality. While time-to-market, cost and performance of devices emerging from these different approaches will determine their use in systems applications, planar integration has distinct advantages that include compactness and reduced packaging costs, and the use of processes that have already been developed for use in silicon integrated circuit technology. Since planar devices use thin-films, the CCS approach to materials investigations can be used to explore new materials for use in planar lightwave circuits (PLC's).

One important function needed in complex WDM optical transmission systems and DWDM (dense wavelength division multiplexed) systems is optical filters. PLC's provide an increasingly important filter technology and have the potential, as noted earlier, for continued integration and increased functionality.

An example of a PLC, an arrayed wavelength grating router (AWG) is shown in Figure 8. The wavelengths that are input on a single fiber are split onto multiple fibers by this device as indicated in the figure. A cross-section of a portion of such a device is also shown. A layer of thermal silica is formed on a (100) silicon wafer. P-doped silica is deposited onto the thermal silicon and patterned into waveguide structures. Finally, a B,P-doped SiO_2 glass is deposited as a capping layer. There is about a 2% difference in index between the glass of the waveguide structures and the glasses used for the surrounding base and cladding layers. The use of higher index materials for the core of the waveguides would allow us to make smaller PLC's. Indeed, high index waveguides are required for future filter designs such as ring resonators.

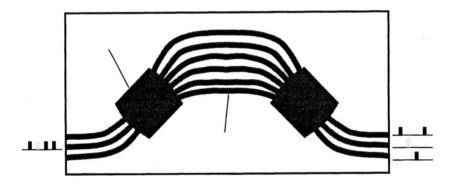

Figure 8. Example of a planar lightwave circuit, an arrayed wavelength router.

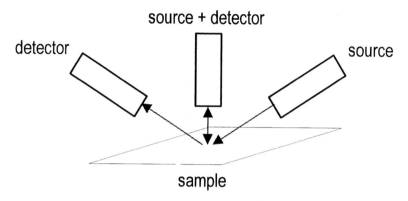

source + detector

detector   source

sample

Figure 9. Schematic representation of the reflectometry instrument.

The CCS approach to optical materials

The materials requirements for PLC's are films that are highly uniform in index and thickness, have low loss in the infared, 1200-1700 nm, and are compatible with existing silicon processing. Moreover, other film deposition techniques, in particular CVD, should be possible for any material identified. While these requirements must be met for device fabrication, the constraints are relaxed during materials investigations. In particular, thickness and index uniformity must be adequate, typically better than a percent, but only over the size of the device to be fabricated.

We have used the CCS approach to examine the addition of metals such as tantalum and zirconium to silica glass to raise the index of refraction. These metals have a single oxidation state, avoiding mixed valence transitions which are fully allowed optical transitions and thus would result in loss. Samples were measure using reflectometry, a technique useful for obtaining index and thickness information on thick films (> 1 μm). Reflectometry is a non-contact mapping technique that measures index and thickness and instruments are commercially available. As shown schematically in Figure 9, the instrument measures reflected white light simultaneously at two different angles. The ratio of the intensities of these reflected beams can be modeled to give index and thickness.

Controlling the tantalum content of the glass controls the index in tantalum-doped silica films. The index change from that of pure silica is indicated in Figure 10. Index as a function of composition for the system $Si_{1-x}Ta_xO_y$ was mapped by reflectometry.

The difference between the index of refraction of pure amorphous silica, a-SiO_2, and the deposited film from which the waveguide structures will be fabricated is known as delta where $\Delta = n_{exp} - n_{SiO2}/n_{SiO2}$. In this system, delta is proportional to the Ta-doping level. In the tantalum doped silica films, the optical quality of the glass looks good and no scattering is observed. However, loss values still must be determined. Also, any issues related to birefringence are unknown but may be addressed by examining the effect of the introduction of additional substituents.

Index @ 632 nm

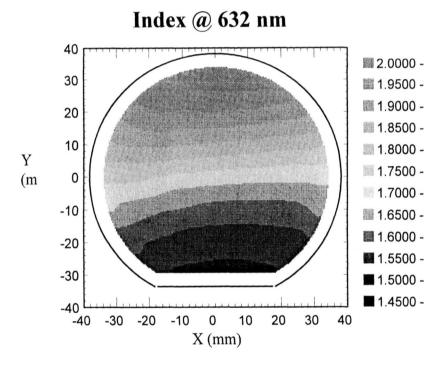

Figure 10. The index as a function of composition for the system $Si_{1-x}Ta_xO_y$.

In principle, a parallel study of the performance of a simple optical test structure could be carried out as depicted in Figure 11. By directly measuring material performance as a function of composition in one experiment, subjects such as processibility, optical loss or materials birefringence could be examined free of run-to-run variations. Such experiments are underway.

Summary

The continuous compositional spread approach to new materials investigations has proved its value by the discovery of a new high specific capacitance, high breakdown field, low loss material in the Zr-Ti-Sn-O phase system. This material may be used for future generations of embedded DRAM structures in integrated circuits. We have also discussed the applicability of the CCS approach to optical materials problems. The use of the CCS approach in the investigation of new high index thin film glasses including the $Si_{1-x}Ta_xO_y$ system for use in planar optical circuits was discussed. More efficient approaches to materials discovery and optimization are likely to be of increasing importance, particularly in the area of optoelectronics.

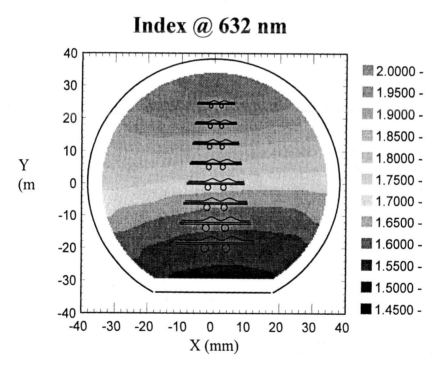

Figure 11. Scheme for a combinatorial-type approach to device characterization.

64

References

* Now with Agere Sytems, Murray Hill, NJ 07974.
1. Dagani, R.; *Chem & Engin. News* **1999**, 77, 51-60.
2. Sawatzky,E; Kay, E; *IBM J. Res. Develop.* **1969**, 13, 696-702.
3. Hanak, J. J.; *J. Mater. Sci.* **1970**, 5, 964-971.
4. Hanak, J. J.; Gittleman, J. I; *Physica* **1971**, 55, 555.
5. Hanak, J. J.; *Jap. J. Appl. Phys. Suppl. 2, Part 1*,**1974,** 809.
6. Xiang, X.-D; Sun, X.; Briceno, G.; Lou, Y.; Wang, K.-A.; Chang, H.; Wallace-Freedman, W. G.; Chen, S.-W.; Schultz, P. G.; *Science*, **1995**, 268, 1738-1740.
7. van Dover, R. B., Schneemeyer, L. F., and Fleming, R. M.; *Nature*, **1998**, 392, 162-164.
8. Gerstenberg, D. In *Handbook of Thin Film Technology;* Maissel, L.; Glang, R.; eds.; McGraw-Hill, NY, 1970, Chapter 19, p 19-8.
9. Hashimoto, C.; Oikawa, H.; Homma, N.; *IEEE Trans. Electron Devices*, **1989**, 36, 14.
10. Iddles, D. M.; Bell, A. J.; Moulson, A. J.; *J. Mater. Sci.* **1992**, 27, 6303-6310.
11. Nakagawara, O.; Toyoda, Y.; Kobayashi, M.; Kayayama, Y.; Tabata, H.; Kawai, T.; *J. Appl. Phys.* **1998**, 80, 388-392.
12. Ramakrishnan, E. S.; Cornett, K. D.; Shapiro, G. H., Howng, W.-Y. *J. Electrochem. Soc.* **1998**, 145, 358-363.
13. White, A. E. *Optics & Photonics News* March, 2000, p. 27-30.
14. Li, Yuan P.; Henry, C. H.; in Optical Fiber Telecommunications IIIB (Academic Press: New York) p.319 (1997).

Chapter 4

Chemically Sensitive High-Throughput Parallel Analysis of Resin Bead-Supported Combinatorial Libraries

J. Lauterbach, C. M. Snively, and G. H. Oskarsdottir

School of Chemical Engineering, Purdue University, 1283 Chemical Engineering Building, West Lafayette, IN 47907-1283

In this study, the feasibility of using Fourier transform infrared (FTIR) imaging for high throughput screening of resin bead supported combinatorial libraries was investigated. FTIR imaging combines the chemical specificity and high sensitivity of infrared spectroscopy with the ability to rapidly analyze multiple samples simultaneously. A new implementation, using a rapid-scan spectrometer instead of a step-scan spectrometer, which had been used in all previous infrared imaging studies, allows much faster collection times without decreasing data quality. Using several model systems, it was established that FTIR imaging is well suited to perform high throughput parallel analysis of ligands supported on polystyrene-based resins. It was found that both parallel identification of selected members of a resin-supported library and parallel studies of *in situ* reaction kinetics on bead supported ligands are possible using the FTIR imaging system. In short, Fourier transform infrared imaging can be employed as a powerful spectroscopic tool for the parallel investigation of combinatorial libraries.

Introduction

For many years, scientists have executed experiments one at a time with very careful control of parameters. Combinatorial chemistry, and high throughput methods in general, suggest that this fundamental concept can be successfully expanded and enhanced. Combinatorial methods were first applied to electronic materials (1). The validation of the combinatorial approach was not immediately achieved, but came about through the application of automated synthesis and high-throughput screening in pharmaceutical research over the past 15 years. In materials science and catalysis, combinatorial methods hold great promise and have attracted a considerable amount of research interest in recent years. The properties of many functional solid-state materials arise from complex interactions that depend heavily on composition and processing conditions. Few general principles have emerged that allow the prediction of structure, synthetic reaction pathways, and resulting properties of such solid-state compounds. The combinatorial process aims at easing this procedure by making it possible to efficiently explore the large parameter space that controls the properties of the final products. In this chapter, we report the application of a novel high-throughput screening method based on spectral imaging in the mid-infrared spectral region (4000 to 900cm^{-1}) to solid support-based combinatorial libraries.

Synthesis of Combinatorial Libraries

The essence of combinatorial synthesis is the ability to generate a large number of chemical species very quickly. In the beginning, the demand for quantity in combinatorial chemistry tended to compromise the quality of the products to some degree, but as combinatorial chemistry secured itself in the scientific world, the quality of combinatorial syntheses improved, and current synthetic methodologies are able to rapidly produce high quality products (2). Many solid phase materials have been developed that can carry organic material and are inert to the conditions of synthesis. Among these are resin beads, multipins, paper or polymer sheets, and even glass chips (2). Resin beads are still the preferred carrier, but multipins are becoming popular in high-throughput pharmaceutical testing. Numerous methods have been set forth to create combinatorial libraries using these solid and liquid phase technique (3). This concept was introduced in 1963 by Merrifield (4). In the 1980s, Geysen reported the parallel synthesis of molecular libraries of peptides in which the peptides were created one amino acid at a time, beginning with an attachment to an array of polyethylene rods (5). By knowing the specific addition history of each pin, the resulting peptide on each pin was known. In another variation of the parallel synthetic scheme, Houghten reported a means of isolating large quantities of different peptides, with permeable "tea-bags" (6). In this scheme, the synthetic histories of individual bags containing many solid supports are tracked as they are subjected to different amino acid additions. The number of fixed supports used to localize the products limits the parallel synthetic route. To increase the number of new molecules, the solid support was miniaturized, giving the submillimeter polymer beads commonly found today. By giving up the requirement of spatial localization of the reactions, millions of potential new molecules could be generated (7). By repeating the so called 'split-and-pool' process of dividing the beads into groups, individually reacting each group in a separate addition step and subsequently recombining all the groups, enormous numbers of different molecules could be made. Although this method is highly efficient at creating broad chemical diversity, the difficulty of identifying the active molecule on each bead still remained. Beginning in the 1990s, methods were developed to encode (tag) individual beads or supports uniquely in order to identify the molecules on individual beads (8).

High-Throughput Screening of Resin Bead Supported Libraries

Methods used to screen the combinatorial libraries are traditionally "one at a time" techniques. These include Fourier transform infrared (FTIR) spectroscopy, solid and liquid phase nuclear magnetic resonance (NMR), and mass spectrometry. Mass spectrometry is very sensitive and can be used to both monitor solid phase products and reactions as well as to identify active compounds from libraries. However, it is a destructive technique, which requires cleavage of compounds being studies from the resin support, and has the disadvantage of requiring an additional chromatographic technique if mixtures are to be analyzed. NMR is an attractive technique because of its nondestructive nature and general easy of sample preparation, and has been used for the analysis of resin-supported ligands (9). Recent efforts have shown promise in the development of high-throughput NMR techniques (10,11), however these remain of limited utility due to the small number of samples that can be simultaneously analyzed and the long analysis times (low sensitivity) inherent to the technique in general.

Infrared-based techniques are extremely popular for the analysis of combinatorial libraries because of the benefits of low cost, ease of use, and rapid data collection (12). FTIR has been used to monitor various solid-phase organic syntheses, where IR spectra can be obtained from samples containing over 100,000 beads by pressing them into KBr pellets (13). This approach, however, has several distinct disadvantages. It destroys the sample, requires rather large amounts of sample, and requires that each bead being analyzed carry the same compound. In libraries manufactured via the "split-and-pool" technique, each bead carries a specific compound. For these libraries, single bead FTIR microspectroscopy has been employed as an analytical tool. This technique uses an FTIR microscope to acquire spectral information from single beads extracted from libraries (14,15). Spectra are acquired from flattened beads in order to minimize optical artifacts and produce spectra with reasonable absorbance intensities. Typical data collection times have been reported to be around one minute per bead (16). While this technique provides highly detailed chemical information, it has the limitation of being able to examine only one bead at a time. It has also been applied to monitor rates of reaction both on the bead surface and in the bead interior (16). To obtain data, beads must be removed from the reaction mixture to be studied under the IR microscope. Therefore, this technique is not completely non-invasive and is not capable of providing truly real-time kinetic information. A recent extension of this technique is FTIR spectroscopic mapping, in which an infrared microscope is used to collect spatially resolved spectra from a collection of beads that are spread out and spatially separated from each other. Using this approach, a spectral map of approximately 300 different resin beads was collected, and the resulting spectra were used to determine the identity of the beads (17). While again providing chemically specific information, this technique requires a collection time of ~5 hours for each map, making real-time kinetic studies impossible.

As a first step towards truly parallel analysis using imaging techniques, infrared thermography has been used to select active supported catalysts (18). By correcting for emissivity differences between materials of interest and the background, thermography can be used to detect activity in combinatorial libraries of catalysts by detecting heat generated from the reaction. Time-resolved IR thermographic detection and IR emission analysis of temperature profiles enable practically any exothermic or endothermic reaction to be monitored in a truly parallel fashion. However, while being able to rapidly analyze a large number of beads at once, this technique provides no chemically specific information, and is therefore of limited usefulness for the analytical characterization of supported ligands. Recently, it has been shown that a new technique employing a near-IR imaging spectrometer can be used to simultaneously monitor the progress of reactions on several different supported resins (19). However, due to the low absorptivity inherent to absorption bands in the near-IR spectral region, this technique requires that the sample contain a large number of supported resin beads in order to generate a measurable signal. Another spectroscopic approach for rapid parallel evaluation of solid bead-based materials is Raman imaging. This technique has been applied to several systems (20). However, it has recently been reported that this technique is of limited utility because the spectra are almost completely dominated by the spectral contributions of the resin support material (21).

Our group has developed what we believe is the first truly parallel, high-throughout, chemically sensitive analytical technique for the analysis of combinatorial libraries (22). This technique is the modern embodiment of spatially resolved infrared spectral imaging.

Infrared Spectral Imaging

Spectral imaging is an interdisciplinary approach that combines chemistry, physics, and engineering. It is used to visualize the chemical composition of materials and is used for research ranging from metabolism in the human brain to formation of the stars (23). The original approach to the collection of spatially resolved infrared information involved the use of a microscope coupled to a dispersive infrared spectrometer (24,25). The next advancement was the incorporation of an FTIR spectrometer with a microscope employing all-reflective optical elements. This technique is a mapping technique, such that spectral information is acquired from discrete regions of the sample, with the resulting information being combined to form a map of that specific region. In order to sample radiation from a well-defined region of the sample, a physical aperture must be placed in the beam path. To collect information from an entire sample, this aperture is translated in a raster-scanning fashion over the desired sampling area. The main drawbacks of this approach are long collection times and limited spatial resolution. The ultimate extension of spatially resolved infrared spectroscopy comes about by the substitution of a focal plane array (FPA) detector for the standard single element detector commonly employed in FTIR instrumentation (26,27). Since this is the key to the entire process, some discussion of this approach is warranted.

Although it is possible to record spectral images with single-point detectors, array detectors greatly expedite image generation because they simultaneously measure light with multiple detector elements. FPAs have been used extensively for remote sensing and astronomical imaging, but have only recently begun to be employed in benchtop scale instrumentation because of the recent increase in availability and price reduction of these devices. The specific detector material determines the spectral region probed. The first detectors to be used for fast imaging applications were made from indium antimonide (InSb). These detectors are limited to the near and near/mid infrared spectral regions (10,000-2,000cm^{-1}). Even though they are not sensitive in the mid-infrared spectral range, they have for example been shown to be able to elucidate differences between phases in mixed water- surfactant system (27). Mercury cadmium telluride (MCT) detectors have been used for fast imaging systems, and are desirable because they detect radiation in the mid-infrared spectral range (4000-900cm^{-1}). Their ability to detect slight differences in chemical structure has been frequently demonstrated.

FTIR imaging has been applied extensively to study biological samples - brain tissue (28), silicone gel in human breast tissues (29), amino acids in a matrix (26), canine bone tissue (30), and pathological and fracture states of bone (31). Applications in the non-biological areas have included the study of heterogeneities in rubber systems (32), solvent diffusion (33) in a polymer film, semi-crystalline polymers (34), and polymer dissolution by mixed solvents (35). Diffusion processes in polymer-liquid crystal systems (36) and polymer-liquid crystal composites have also been examined (37).

What these previous studies have in common is that they all imaged a single sample. The idea here is that instead of examining multiple spatial regions of a single sample, multiple samples are analyzed simultaneously, as illustrated in Figure 1. Our group has pioneered FTIR imaging to perform simultaneous measurements of the IR absorption by molecular vibrations in multiple members of a combinatorial library, allowing the *in situ* parallel investigation of chemical processes.

Practical Aspects of Spectral Imaging

The FPA and the FTIR spectrometer are synchronized such that the modulation of the source or signal by the interferometer is synchronized with image collection by the FPA. In operation, a series of images is collected as a function of interferometer optical path difference. One way to do this, which has been the preferred method to date, is to operate the interferometer in a step-scan mode to collect a series of spatially-resolved interferograms. These interferograms are then converted to the frequency domain by performing a Fourier transform for each pixel. This way, IR spectra are recorded for every spatial location in the image plane simultaneously. The data can therefore be thought of as a three-dimensional "image cube" that spans one spectral and two spatial dimensions. The data can be manipulated either as a series of wavelength-resolved images or alternatively as a series of spatially resolved spectra, one for each point in the image. The spectral parameters (resolution, wavelength range) are determined by the standard rules for FTIR spectroscopy (38), and the spatial resolution depends on the optical configuration and varies with spectral range (maximum diffraction limited resolution of ~3μm using an InSb detector and ~7μm using an MCT detector). Instrumental modifications made in our laboratory (39) have allowed us to improve this scheme to utilize a rapid-scanning interferometer.

As stated above, imaging data can be manipulated in either the spectral or spatial domain. In the spectral domain, the spectrum of a specific pixel in utilized to determine the chemical nature of that particular region of the sample. In the spatial domain, the absorbance value of a particular absorption band is plotted for each pixel to generate a spatially resolved, chemically specific image. In the spectral domain, all of the processing routines commonly encountered in FTIR data processing, such as baseline correction, integration, peak rationing, and truncation, can be performed. In the spatial domain, standard image analysis

71

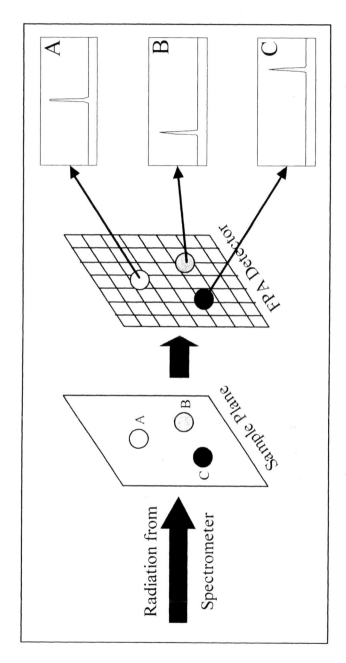

Figure 1: Illustration of the basic concept of the application of FTIR imaging to high-throughput screening of resin supported libraries. The throughput of this approach is directly proportional to the number of samples that can be placed in the field of view of the instrument.

and enhancement routines can be used to manipulate the data. Smoothing and sharpening operations can be performed, and information on the size, shape, and area of specific features in an image can be determined.

Due to the dual spatial/spectral nature of the data collected in these experiments, additional manipulations are possible. A commonly used example of this is the generation of band ratio images, in which the intensity value at each pixel position is the result of the ratio of intensities of two distinct spectral bands from two images. This is commonly used either as a thickness correction method or as a way to determine the relative ratio of two components in different regions of a sample. Additionally, the same band from different images of the same sample can be ratioed to form a unique image. This approach has been used to generate dichroic ratio images (34). Recent interest (40) has focused on the application of multivariate techniques to the analysis of imaging data sets . These techniques are particularly necessary in the analysis of spectral images due to the increased amount of data present compared to standard infrared spectroscopic experiments.

Experimental Setup

Rapid Scan Imaging Setup

In all previous studies, FTIR imaging instrumentation consisted of a step-scan FTIR spectrometer as a light source, a microscope with Cassegrainian optics, and a focal-plane array (FPA) detector. When a step scan spectrometer is used to collect a data set, the moving mirror of the interferometer is stopped at specified intervals of retardation, and the light intensity values at all pixels (one image frame) are measured. Several frames are collected at each interferogram data point and averaged to obtain a reasonable signal-to-noise ratio (SNR). This collection process requires that the retardation of the interferometer remain constant during data collection. Therefore, the overall collection time of a single data set is 3-15 min on average, depending on the desired spectral resolution and mirror velocity.

When a rapid scan spectrometer is used to collect data, the light intensity values for all the pixels are measured at specific times, while the mirror is continuously moving. This instrumental approach was recently reported for the first time (39). For a rapid-scan spectrometer, each data point is collected over a period of a few hundred microseconds while the interferometer mirror is moving. Therefore, the data at each point is spread out over a range of retardations. Two instrumental modifications allow us to circumvent this problem and to employ successfully a rapid-scan spectrometer as the light source for the mid-infrared image. First, the electronics employed here are sensitive enough to allow the collection of data with a reasonable signal to noise ratio by acquisition of only one frame per interferogram data point. Additionally, since we are acquiring only a single frame at each data point, any artifacts that would result from collecting data over a range of retardation are minimized. We use a frame rate of 180 Hz and an integration time of 180μs. With a mirror velocity of 0.0158 cms^{-1}, data are collected over a retardation of 28 nm at each data point, which compares reasonably with the positional accuracy of an interferometer in step-scan mode. Moreover, owing to the decreased collection time of this technique, it becomes practical to average data from several experiments together to increase the SNR, which is common practice in non-imaging FTIR spectroscopy. This averaging technique has been shown to be superior to the technique of collecting multiple frames at each retardation (41). Overall, the use of a rapid-scan spectrometer has the advantage of reducing the data-collection time by at least an order of magnitude. A data set with 8cm^{-1} spectral resolution over 4000cm^{-1} spectral range can be collected in just under 9 min by use of the conventional step-scan technique. The rapid-scan experimental setup allows us to collect a data set with 8cm^{-1} resolution over a 1360cm^{-1} spectral range in ~17s. It should be noted here that the incorporation of this new rapid scan technique achieves a dramatically reduced collection time without loss of data quality. It has been shown that, for a specific sample geometry, the rapid scan setup gives a SNR of 98, whereas the step scan technique gives a SNR of 95 (39). When the collection time of each of these techniques is taken into consideration, the rapid scan technique represents approximately a 30-fold increase in SNR.

The optical setup consists of a Nicolet Magna 860 FTIR spectrometer, BaF$_2$ condensing and refocusing optics, a wide bandpass filter, a KBr diffuser, and a 64×64 pixel mercury cadmium telluride FPA detector (39). To perform a rapid-scan experiment, the spectrometer scanning speed and resolution are set. The forward movement of the interferometer mirror triggers data collection. A computer then collects the maximum number of data points based on the scan speed, spectral resolution, and the FPA frame rate. A bandpass filter is employed to allow only light from a specific spectral region to be collected and to minimize Fourier fold-over noise (38). In a single experiment, a 64×64×1360 point data set is collected, occupying approximately 11 MB of disk space.

Sample Cell Design for Resin Bead Studies

In order to study resin bead libraries, IR transmission sampling was used. The sample holder was made of two calcium fluoride (CaF$_2$) windows and two variable path length metal frames that held the windows together. To minimize the stress on the windows when the frames were tightened, rubber O-rings were placed between the frames and the windows. In order to effectively investigate beads using this technique, the beads should be flattened to the desired thickness. This is accomplished by introducing solvent, which causes the beads to swell. Once swollen, the beads are flattened to the desired thickness by application of slight pressure. Additionally, analyzing swollen beads has the advantage that optical artifacts are minimized when the solvent is present. A sharp change in refractive index across a boundary, such as that between a flattened bead and air, causes a thick, dark ring to appear due to a lensing effect. This effect is minimized by surrounding the bead with a medium of higher refractive index (many organic solvents have refractive index ~1.5) than that of air (refractive index very close to 1.0). Both the lensing and swelling effects upon introduction of solvent can be seen in figure 2.

The first type of experiment was the identification of specific amino acid-carrying beads in a mixture of four different amino acid-carrying beads. The beads were placed onto a polished CaF_2 window and swollen with methylene chloride. The solvent was allowed to partially evaporate, and another CaF_2 window was placed on top. Slight pressure was maintained in order to flatten the beads, which was necessary to obtain spectra free from saturation effects. This sample was placed in the field of view of the spectrometer, and images were collected. After the analysis, the beads could be recovered without damage and submitted for further analysis by re-swelling them in solvent, causing them to resume their spherical shape.

The second class of experimentation was tailored to examine specific groups of beads placed in the imaging spectrometer in a systematic fashion. This approach is particularly amenable to studies of reaction rates by removing beads from a reaction mixture at different times during the reaction and then analyzing all groups together in a single experiment. The sampling geometry was very similar to the one used for the identification studies. The only difference was that when arranging the beads on the window, a rectangular plastic grid was used to help arrange the groups and to prevent mixing among beads removed from the reaction mixture at different times.

For *in situ* kinetic studies of reactions, a flow cell was designed and employed. The flow cell was composed of two BaF_2 plates separated by a Teflon spacer. The top plate had two holes machines into it to allow the introduction and removal of solvent. The thickness of the Teflon spacer (typically 50μm) was selected to optimize the measured infrared absorbance intensity. Several beads were placed on the bottom plate, a spacer was added, and the top plate was placed on top. Pressure was applied to the cell to ensure a leak-tight seal. The flow of solvent into the flow cell was controlled using a syringe directly connected to the input of the flow cell.

For all of these experiments, the increase in throughput brought about by using FTIR imaging instead of standard single bead FTIR microspectroscopy is directly proportional to the number of samples that can be placed in the field of view of the instrument.

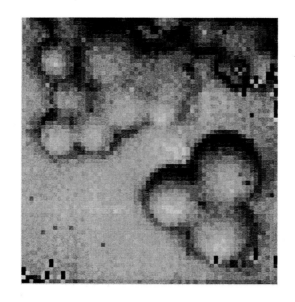

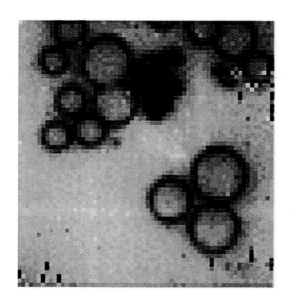

Figure2: Beads flattened between CaF$_2$ plates before introduction of solvent (left) and after introduction of solvent (right)

Screening of Resin Supported Libraries

The screening and structural characterization phases have consistently been the bottleneck in the overall process and still constitute in most cases a very challenging task. As discussed before, many of the existing methods are not capable of non-invasive, rapid, high throughput, and real-time spectroscopic monitoring of a binding or catalytic event within or around a resin-bound ligand or catalyst. We will use three examples to demonstrate that FTIR spectral imaging is capable of monitoring not only the kinetics of the reaction on multiple resin beads, but also provides information about the reaction mechanism and selectivity that is simply not available using other techniques.

Identification of Beads Carrying Different Types of Ligands

Three popular supports used in solid phase organic synthesis are (1) crosslinked (1%-divinylbenzene) polystyrene resins, (2) polystyrene-based resin extensively grafted with polyethylene glycol linkers (PS-PEG) and (3) the surface-modified polyethylene pins. Split and pool synthesis of libraries is a fast and effective way to create beads with different types of ligands attached. To recognize all the different ligands that have been synthesized has until now required some type of invasive method such as tagging the supports with chemical entities that can more easily be recognized. The use of FTIR imaging to identify individual ligands in a group of beads makes such treatment unnecessary.

To investigate the feasibility of FTIR imaging for high throughput analysis of resin bead libraries, a mixture of about 25 beads comprised of four different ligands was analyzed. The ligands were protected amino acids attached to Merrifield Resins, purchased from Advanced Chemtech. Using FTIR imaging allows a much faster and higher throughput analysis of the data. Instead of having to identify each bead through its IR spectrum, one can create a chemically specific image that permits all the beads in the mixture to be categorized into one of the four groups. The images from this experiment are shown in Figure 3. By selecting a spectral frequency that is unique for each ligand and plotting the absorbance value for each pixel at that frequency, an image is generated that clearly reveals the location of each type of supported ligand. It is of course also easily possible to identify beads according to specific properties of their support material. Principal component analysis (42) can also be interfaced with this technique to identify ligands either where the spectra are highly convoluted (such as those containing a longer sequence of amino acid residues) or where a minimum of human intervention is desired (43).

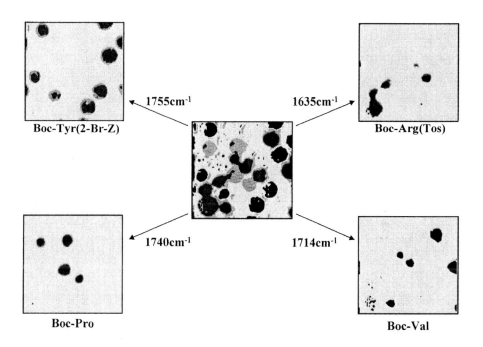

Boc-Tyr(2-Br-Z)   1755cm⁻¹   1635cm⁻¹   Boc-Arg(Tos)

Boc-Pro   1740cm⁻¹   1714cm⁻¹   Boc-Val

*Figure 3: Images of a mixture of beads with supported amino acid ligands.
Images were generated by plotting bands specific to Boc-Pro (C=O stretch,
1740 cm⁻¹), Boc-Tyr(2-Br-Z) (C=O stretch, 1755 cm⁻¹), Boc-Arg(Tos) (C=N
stretch, 1635 cm⁻¹), and Boc-Val (C=O stretch, 1714 cm⁻¹). Spectral images
were generated by plotting bands specific to each ligand, showing the spatial
location of each type of bead. The center image is a composite of all images
(Reprinted with permission from (22). Copyright 2000 American Chemical
Society).*

Ex situ Reaction Kinetic Studies on Resin Beads

The second geometry that was studied was a resin bead library generated from well-separated groups of beads where characteristics of each group were to be investigated in parallel. The different groups of beads were isolated from a reaction mixture at different times during the course of the reaction and were analyzed simultaneously. The goal was to greatly reduce the time required to collect kinetic data, by using the parallel analysis capabilities of FTIR imaging. The example demonstrated here is the oxidation of a primary alcohol to an aldehyde, which has been described previously in the literature (44,45). 117.4 mg 4-Methylmorpholine N-oxide (NMO), 97%, Aldrich and 7.1mg tetrapropylammonium perruthenate (TPAP), Aldrich were dissolved in 5ml dimethylformamide (DMF), Aldrich. About 50 mg of beads were placed in the solution, and the reaction mixture was continuously stirred. At specific time intervals, about 50 beads were removed from the reaction mixture, and the reaction was quenched by washing the beads with tetrahydrofuran (THF) and methylene chloride. Subsequently, while still in the methylene chloride solvent a portion of the beads was removed and placed onto a polished CaF_2 window. The beads removed at different times were arranged into groups, and with the help of a previously described plastic form, each group was well separated from the other (Figure 4). As in the previous example, after the solvent had mostly evaporated, the plastic form was removed, another CaF_2 window was placed on top, and the beads were imaged.

The data from this experiment are shown in Figure 5. The image at the top was generated using the raw output from the FPA, which is averaged over all collected wavelengths. Spectra from each group of beads in the spectral imaging data set were then used to quantify the kinetics of the reaction. During the reaction, primary alcohol is converted to an aldehyde, and the C=O stretching band at $1688 cm^{-1}$ increases in intensity. This band was used to quantify the conversion for this reaction. Thickness correction was performed by dividing the integrated peak area of the $1688 cm^{-1}$ band by the integrated peak area of the aromatic combination band around $1945 cm^{-1}$, which corresponds to the polystyrene support and should be unaffected by the reaction. The rate constant for the reaction was estimated by plotting the resulting thickness corrected band from each group of beads versus the time when the beads were removed from the reaction mixture (i.e. reaction time). The conversion of alcohol to aldehyde fits well to an exponential expression of the type $y = y_0 - A_1 exp(-k*t)$ (44). From a nonlinear curve fitting routine, we obtain a first order reaction rate constant of k $= (3.5 \pm 0.6) \times 10^{-4} s^{-1}$. This value agrees within experimental error to values determined in previous studies (44,45). The error for the kinetic plot was determined using the method of least squares.

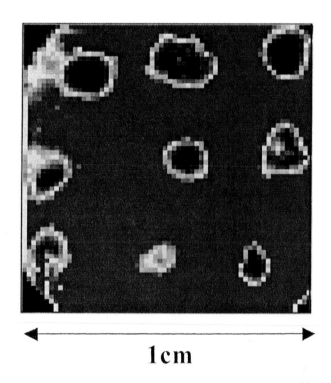

1cm

Figure 4. Image of groups of beads, each removed at different times during the reaction described in the text. Each small group consists of approximately 10-20 beads (Reprinted with permission from (22). Copyright 2000 American Chemical Society).

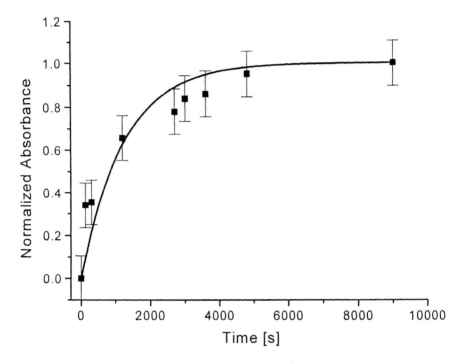

Figure 5: Plot of the thickness corrected 1688 cm^{-1} C=O stretching band from spectra corresponding to each collection of beads in Fig. 4.The error bars were determined from the peak-to-peak noise of the absorbance spectra.

In situ Reaction Kinetic Studies on Resin Beads

Due to the dramatically decreased collection time of rapid scan FTIR imaging, it becomes possible to acquire *in situ* kinetic information from much faster processes. The ability to examine the progress of reactions *in situ* has several advantages. The reaction system does not need to be perturbed in order to be analyzed, so it is no longer necessary to remove beads from the reaction mixture and quench the reaction. This removes one step from the process and also decreases the chances of introducing additional contaminants into the system. This method also has the advantage of ease of use. Once the beads are placed within the flow cell and the reaction is started, the only thing that remains to be done is data collection. There is no need to repetitively manipulate and analyze single beads, such as in single bead FTIR microspectroscopy. Only a single sample preparation step is required for this analysis, which not only reduces the amount of time and effort placed into the analysis, but also lessens the chance of introducing contaminants into the system.

As a demonstration of the utility of this technique, the same reaction as described in the above section was carried out in a flow cell. However, in this experiment, the amount of catalyst was increased in order to increase the reaction rate to truly test the ability to follow faster processes. The absorbance intensity of the aldehyde carbonyl band from one bead over the course of the reaction is plotted in figure 6. The fit to the data gives a first order reaction constant of $k = (3.2 +/- 0.9) \times 10^{-3}$ s^{-1}, which is faster than in the previous study, as expected. Several other beads were examined, and the rate constants determined from all beads examined was found to be the same, within experimental error. It should be noted here that, if step scan FTIR imaging was used to acquire similar data, a maximum of about six data points could have been collected in the same period of time, which would have imposed a severe limitation on the quality of the fit, and on the resulting quantitative information.

One point to be aware of is that, in order to effectively compare the results between beads, all beads examined must be subjected to as nearly the same reaction conditions as possible. Since these reactions were carried out in a flow cell, both the access to fresh reactants and the removal of products is dictated by the nature of the flow in and around each bead. Ideally, one would like to have all of the beads arranged in a straight line, with the flow perpendicular to this line. This would assure that each bead is seeing very nearly the same flow, and that none of the beads are receiving less solvent because they are "hidden" from the flow behind other beads. This effect is noticeable in the example shown in figure 2. In this case, the flow was from the top of the image to the bottom. Therefore, beads that were behind other beads demonstrated a much reduced reaction rate, and therefore a lower total conversion on the time scale of the experiment. The data mentioned in the above paragraph were taken from beads on

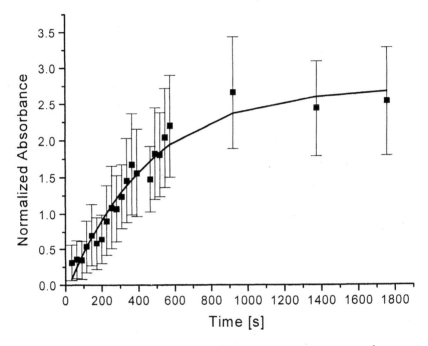

Figure 6: Representative plot of the thickness corrected 1688 cm^{-1} C=O stretching band from spectra collected in situ during reaction from one of the beads shown in figure 2.

the periphery of the collection of beads, which should experience the maximum of solvent flow.

Summary

We have used FTIR imaging as a spectroscopic tool for the parallel identification of selected members of resin-supported combinatorial libraries. This technique combines the chemical specificity and high sensitivity of mid-IR (4000-1000 cm^{-1}) imaging with the ability to rapidly analyze multiple samples simultaneously. This includes both the identification of supported ligands and also the monitoring of reactions occurring on supported ligands. The chemical identity of a variety of ligands supported on resin beads can be determined in a single experiment without perturbing the system.

In order to apply FTIR imaging to longer peptide sequences, more complex supported ligands, or ligands that have similar chemical structures, more complex types of classification routines, such as multivariate techniques (42), must be interfaced with this technique. The application of multivariate techniques to spectral imaging data will not only increase the flexibility of spectral imaging, but will additionally allow the vast quantities of data obtained to be reduced to a more usable form.

In summary, FTIR spectral imaging poses an elegant solution for the rapid characterization of resin-supported library members without the need for chemical encoding methods or other invasive analytical techniques.

Acknowledgements

This work was supported by the National Science Foundation (CTS-9871020 and CTS-0071020).

References

(1)   Hanak, J. J. *J. Mat. Ssci.* **1970**, *5*, 964-971.
(2)   Terrett, N. K. *Combinatorial Chemistry*; Oxford University Press Inc.: New York, 1998.
(3)   Czarnik, A. W.; DeWitt, S. H. *A Practical Guide to Combinatorial Chemistry*; American Chemical Society: Washington, DC, 1997.
(4)   Merrifield, R. B. *J. Am. Chem. Soc.* **1963**, *85*, 2149-2154.
(5)   Geysen, H. M.; Meloen, R. H.; Barteling, S. J. *Proc. Nat. Acad. Sci.* **1984**, *81*, 3998-4002.
(6)   Houghten, R. A. *Proc. Nat. Acad. Sci.* **1985**, *82*, 5131-5135.

84

(7)   Furka, A.; Sebestyen, F.; Asgedom, M.; Dibo, G. *Int. J. Pep. Prot. Res.*
        1991, *37*, 487-493.
(8)   Brenner, S.; Lerner, R. A. *Proc. Nat. Acad. Sci.* **1992**, *89*, 5381-5383.
(9)   Shapiro, M. J.; Chin, J.; Marti, R. E.; Jarosinski, M. A. *Tetrahedron
        Letters* **1997**, *38*, 1333-1336.
(10)   Hou, T.; MacNamara, E.; Raftery, D. *Anal. Chim. Acta* **1999**, *400*, 297-
        305.
(11)   Subramanian, R.; Webb, A. G. *Anal. Chem.* **1998**, *70*, 2454-2458.
(12)   Yan, B.; Gremlich, H.-U.; Moss, S.; Coppola, G. M.; Sun, Q.; Liu, L. *J.
        Combi. Chem.* **1999**, *1*, 46-54.
(13)   Chen, C. X.; Randall, L. A. A.; Miller, R. B.; Jones, A. D.; Kurth, M. J.
        J. Am. Chem. Soc. **1994**, *116*, 2661-2662.
(14)   Shapiro, M. J.; Lin, M.; Yan, B. In *A Practical Guide to Combinatorial
        Chemistry*; Czarnik, A. W., DeWitt, S. H., Eds.; American Chmical
        Society: Washington, 1997; pp 123-151.
(15)   Yan, B.; Kumaravel, G. *Tetrahedron* **1996**, *52*, 843-848.
(16)   Yan, B.; Fell, J. B.; Kumaravel, G. *J. Org. Chem.* **1996**, *61*, 7467-7472.
(17)   Haap, W. J.; Walk, T. B.; Jung, G. *Angew. Chem. Int. Ed.* **1998**, *37*,
        3311-3314.
(18)   Taylor, S. J.; Morken, J. P. *Science* **1998**, *280*, 267-270.
(19)   Fischer, M.; Tran, C. D. *Anal. Chem.* **1999**, *71*, 2255-2261.
(20)   Pivonka, D. E. *J. Combi. Chem.* **2000**, *2*, 33-38.
(21)   Fenniri, H.; Achkar, J.; Mathivanan, P.; Hedderich, H. G.; BenAmotz,
        D.; Snively, C. M.; Katzenberger, S.; Oskarsdottir, G.; Lauterbach, J.
        Abstr. Pap. Am. Chem. Soc. **2000**, *219*, 467.
(22)   Snively, C. M.; Oskarsdottir, G.; J., L. *J. Combi. Chem.* **2000**, *2*, 243-
        245.
(23)   Colarusso, P.; Kidder, L. H.; Levin, I. W.; Fraser, J. C.; Arens, J. F.;
        Lewis, E. N.; *Appl. Spec.* **1998**, *52*, 106A-120A.
(24)   Gore, R. C. *Science* **1949**, *110*, 710-712.
(25)   Coates, V. J.; Offner, A.; Siegler, E. H. *J. Opt. Soc. Am.* **1953**, *43*, 984-
        991.
(26)   Lewis, E. N.; Levin, I. W. *Appl. Spec.* **1995**, *49*, 672-678.
(27)   Lewis, E. N.; Treado, P. J.; Reeder, R. C.; Story, G. M.; Dowrey, A. E.;
        Marcott, C.; Levin, I. W. *Anal. Chem.* **1995**, *67*, 3377-3381.
(28)   Lewis, E. N.; Gorbach, A. M.; Marcott, C.; Levin, I. W. *Applied
        Spectroscopy* **1996**, *50*, 263-269.
(29)   Kidder, L. H.; Klasinsky, V. F.; Luke, J. L.; Levin, I. W.; Lewis, E. N.
        Nat. Med. **1997**, *3*, 235-237.
(30)   Marcott, C.; Reeder, R. C.; Paschalis, E. P.; Tatakis, D. N.; Boskey, A.
        L.; Mendelsohn, R. *Cell. Mol. Biol.* **1998**, *44*, 109-115.
(31)   Mendelsohn, R.; Paschalis, E. P.; Sherman, P. J.; Boskey, A. L. *Appl.
        Spec.* **2000**, *54*, 1183-1191.
(32)   Oh, S. J.; Koenig, J. L. *Anal. Chem.* **1998**, *70*, 1768-1772.

(33)   Snively, C. M.; Koenig, J. L. *Macromolecules* **1998**, *31*, 3753-3755.

(34)   Snively, C. M.; Koenig, J. L. *J. of Polym. Sci.* **1999**, 2353-2359.

(35)   Bhargava, R.; Ribar, T.; Koenig, J. L. *Appl. Spec.* **1999**, *53*, 1313-1322.

(36)   Snively, C. M.; Koenig, J. L. *J. of Polym. Sci.* **1999**, *37*, 2261-2268.

(37)   Bhargava, R.; Wang, S. Q.; Koenig, J. L. *Macrmol.* **1999**, *32*, 8982-8988.

(38)   Griffiths, P. R.; de Haseth, J. A. *Fourier Transform Infrared Spectrometry*; Wiley: New York, 1986.

(39)   Snively, C. M.; Katzenberger, S.; Oskarsdottir, G.; Lauterbach, J. *Opt. Lett.* **1999**, *24*, 1841-1843.

(40)   Budevska, B. O. *Vib. Spectros.* **2000**, *24*, 37-45.

(41)   Snively, C. M.; Koenig, J. L. *Appl. Spec.* **1999**, *53*, 170-177.

(42)   Koenig, J. L. *Spectroscopy of Polymers*; American Chemical Society: Washington, DC, 1992.

(43)   Tran, C.; Alexander, T. In *FACSS*: Nashville, TN, 2000.

(44)   Yan, B.; Sun, Q.; Wareing, J. R.; Jewell, C. F. *J. Org. Chem.* **1996**, *61*, 8765-8770.

(45)   Li, W.; Yan, B. *J. Org. Chem.* **1998**, *63*, 4092-4097.

Chapter 5

Parallel Pressure Reactor System

P. Wright, J. Wasson, F. Gebrehiwet, Y. Yun, J. Labadie, T. Long, and G. S. Hsiao

Argonaut Technologies, 887 Industrial Road, Suite G, San Carlos, CA 94070

A system for the acceleration of catalyst screening is described. This system incorporates multiple independent reactors operating in parallel to accelerate the screening process. The instrument is designed for screening pressured reactions with a gaseous reactant. Details of the design of this instrument, together with chemistry obtained on the instrument are presented.

The use of catalysts for pressurized reactions such as hydrogenations and olefin polymerizations is widespread. These processes are of great economic important, and the use of hydrogenations[1] is critical in the manufacturing of pharmaceutical drugs. Improvements in the efficiencies of such processes can make new and novel plastics readily available and accelerate the discovery of new drugs.

To optimize such a process an enormous number of variables must be screened. These include various catalysts, solvent, temperatures, and pressures. The complexity of this task has prompted catalyst manufacturers to provide guides for selecting the appropriate conditions for a particular reaction[2]. Until recently the equipment used for such screening has remained essentially unchanged over many decades.

The principles and concepts used in high throughput screening for biological assays are being increasingly applied to the field of chemistry. The design and functionality of a new instrument[3] (Endeavor™) for catalyst screening is described in this paper. Examples of engineering performance and actual catalyst screens are presented.

Instrumentation

Hardware

Reactors

The system uses 8 reactors made of 316 stainless steel. These reactors are lined edge-on-edge across the system. This minimizes the footprint of the instrument, a key requirement for operation in a fume hood or dry box. A photograph of the instrument is shown in Figure 1. Each of the reactors operates in an independent fashion. A cut-away of a reactor is shown in Figure 2. The thickness of the walls can withstand a theoretical operating pressure of nearly 200 bar, although the actual operating pressure is limited to 35 bar.

At either side of the reactor enclosure there is a thermocouple and heater cartridge. The thermocouple was placed inside the reactor block rather than in the actual reaction solution[4]. This minimizes the need for cleaning, a frequent concern for many chemical reactions, especially olefin polymerization. The inside of the reactor contains a disposable glass liner, which contains the liquid chemicals used in the reaction.

Gas enters in the headspace above the glass liner from a port on the backside of the reactor. Liquids can be introduced into the reactor prior to sealing or through a check valve and into tubing which empties into the glass liner. The

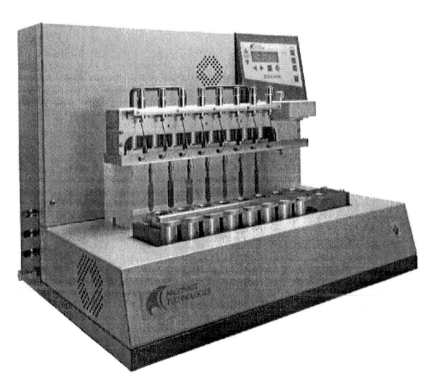

Figure 1-Photograph of the parallel pressure reactor system. Maximum overall dimensions are 56 cm in length, 44 cm in height, and 41 cm in depth.

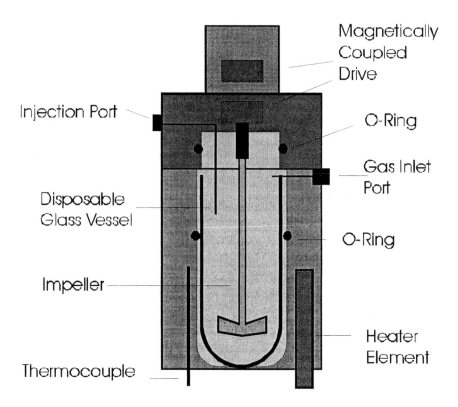

Figure 2-Cut-away diagram of an individual reactor showing the various elements including thermocouple, heater cartridge, impeller shaft, top seal, injector port, and glass liner.

reactors are all sealed simultaneously by securing the stirrer assembly with a pair of clamps. Symmetric tightening of the clamps ensures a pressure tight seal. The clamps grip the stirrer assembly from the top and the reactors from the bottom. This design continues to hold the two components together even when the bolts are loosened under pressure, thereby preventing accidentally injury.

Impellers

The impellers stir the liquid contents of the glass liner and facilitate the dissolution of gas into solution. The dissolution of gas into solution, known as mass transfer, is a critical step in the overall kinetics of a gaseous reaction such as hydrogenation[5]. A special impeller design was created to facilitate mass transfer. Instead of the conventional gas entrainment impeller, a broad large displacement paddle created an unstable vortex which facilitated gas dissolution. All eight impellers are driven in a daisy chain fashion by a motor allowing speeds of up to 1000 rpm. Magnetic couplers are use to ensure gas tight seals and prevent motor damage when working with high viscosity solutions such as those encountered in olefin polymerization. Finally the impellers are made of injection molded PEEK and have a quick disconnect to facilitate ready replacement and disposal.

Electronics

The electronics of the instrument are encompassed in the overall housing to minimize the footprint. For safety concerns the housing is continuously ventilated, all motors are brush less, and solid state electronics are used to eliminate any possible spark sources.

Temperature Control

A feedback loop is used to control the temperature of each individual reactor. Since each reactor has its own unique thermal environment, the control algorithm for each reactor is different. The temperature is calibrated against the temperature measured inside the actual solution in the glass liner where a reaction occurs. The large mass of the reactor block provides temperature

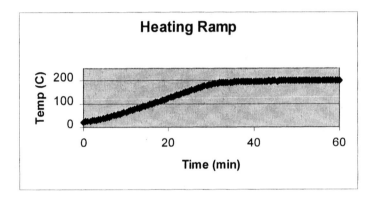

Figure 3- Graph showing the temperature as a function of time for an individual reactor as it heats from ambient to 200° C.

stability for exothermic reactions and smooth ramp up of the temperature to the maximum[6] set value of 200° C as shown in Figure 3.

Pressure Control

The pressure control algorithm requires both smooth ramp up of the pressure to the set value, along with a regulation at that value to replenish gas consumed in a reaction. Because of the small gas volume (~15 ml) of the reactors pressurization occurs in incremental steps. The reactors are connected to a common manifold with individual solenoid valves and piezoelectric pressure transducers for each reactor as shown in Figure 4. Brief pulses are used to open the valves for increments of a few milliseconds. The pressure is measured after the valve is closed. If little or no pressure increase is detected, the valve opening pulse is lengthened. This adaptive algorithm accounts for differences between valves and the pressure gradient between reactors and manifold. This is key for mass manufactured instrumentation implemented in a range of operating environments. As the measured pressure approaches the set value, the valve opening pulse is again shortened to prevent overshooting the desired pressure. The effectiveness of this algorithm in pressurizing all reactors simultaneously to different pressures from a single manifold is demonstrated in Figure 5.

Software

Programming

The instrument can either be controlled from a PC using software or directly from a keypad built into the instrument. Temperatures, pressures, and reaction times can be set for each individual reactor. The opening and closing of valves to purge the system with inert gas is controlled under a single command where the number of cycles is specified. Software control allows more flexibility in the sequence of operations as might be required for some types of chemistry. For example, some chemical transformations involve sequential pressurized reactions such as hydrogenation and carbonylation. However, keypad control is sufficient for >90% of the pressurized chemical transformations practiced in fine chemicals such as batch hydrogenations and carbonylation.

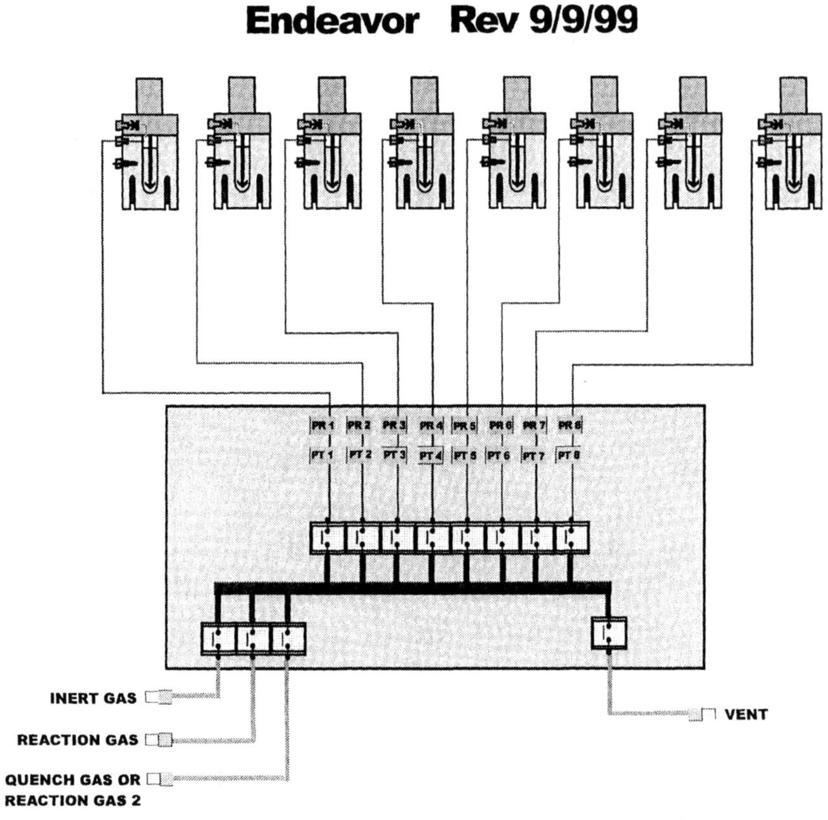

Endeavor   Rev 9/9/99

Figure 4- Schematic diagram showing the gas lines contained within the instrument. Three gases feed into a common manifold, each reactor has its own pressure transducer and valve to regulate the pressure within that reactor.

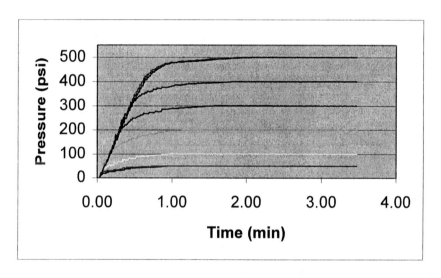

Figure 5- Graph showing ramping of the pressure in a single experiment. Set values were 50 (2x), 100, 200, 300, 400, 500 (2x) psi. Note the consistency for duplicate pressure settings and the critical damping of the pressure ramping algorithm that prevents overshooting the set value

Data Acquisition

During the course of a reaction data are recorded for all reactors. These include the temperature, pressure, and gas consumption associated with each individual reactor. The gas consumption is calculated using the ideal gas law

$$PV=nRT \qquad (1)$$

This requires the user input the volume of solution used in each reactor. The gas consumption measurements are initiated once the set temperature and pressure are reached. Users report the typical measured gas consumption is within 5-15% of the theoretical value. The limitations of the ideal gas law, solubility and vapor pressure effects, and the presence of gas generating side reactions contribute to the deviation from the theoretical value.

Analysis

Data are displayed in a graph on a connected PC in real time. The data from one, many, or all reactors are be displayed in a standard 2D color graph. Temperature, pressure, and /or gas consumption are plotted vs. time for visual analysis. The data are saved in an ASCII format; further numerical analysis is possible using standard commercially available software.

Chemistry

Quantitative Hydrogenation

After the engineering performance tests of the prototype were complete, chemistry validation of the instrument was undertaken. The hydrogenation of a class of pharmacologically relevant molecules was undertaken. 3-substituted indolin-2-ones have been found to have important biological activity as ATP-site inhibitors of Receptor Tyrosine Kinases (RTKs)[7].

Experimental

Compounds **1a-h** (Figure 6) were prepared by analogy to the reported procedure with a Quest™ 205 using oxindole[8], aromatic aldehydes (1.2 equiv), and piperidine (10 mole %) on a 10 mmole scale. To the glass liner was added **1** (1.0 mmol), 10 wt. % catalyst palladium or platinum on carbon (10 wt.% relative to **1**), and 5 mL THF and the liner was inserted into the pressure reactors. For substrates **1a-d** catalytic hydrogenation was performed with 10 wt.% Pd/C. For substrates **1e-h**, 10 wt.% Pt/C was used to avoid dehalogenation[1]. The hydrogen pressure was chosen to be 4 bar, based on conditions reported for trisubstituted olefins[1,9]. The instrument was programmed to purge with hydrogen and pressurize to 4 bar at 50 ° C with a runtime of 6 h. All reagents and catalysts were obtained from Aldrich Chemical Company.

The reaction mixtures were filtered through fine filter paper and concentrated to afford the compounds **2a-h** (Figure 7). The product was weighed and analyzed by HPLC (Alltech Rocket column C18 3 μm column) $CH_3:H_2O$ with 0.1 % TFA, 10 – 100 (4.5 min), 254 nm, and 1 H NMR. Representative 1H NMR spectral data is provided for **1a** and **2a**:

1a: 1 H NMR ($CDCl_3$, 300 MHz): d 7.74 (d, J = 7.8 Hz, 1H, CH), 7.70 (bs, 1H, NH), 7.68 (bs, 1H, CH), 7.25 – 7.17 (m, 3H, CH), 6.92 – 6.84 (m, 3H, CH), 6.05 (s, 2H, CH_2) ppm.

2a: 1 H NMR ($CDCl_3$, 300 MHz): d 8.97 (bs, 1H, NH), 7.17 (t, J= 7.5 Hz, 1H, CH), 6.92 (t, J = 7.5 Hz, 1H, CH), 6.84 (t, J = 9.0 Hz, 2H, CH), 6.68 – 6.66 (m, 2H, CH), 6.60 (dd, J = 6.3, 8.1 Hz, 1H, CH), 5.90 (s, 2H, CH_2), 3.68 (dd, J = 4.5, 8.7 Hz, 1H, CH_2), 3.39 (dd, J = 4.8, 13.8 Hz, 1H, CH_2), 2.88 (dd, J = 9.0, 14.1 Hz, 1H, CH) ppm.

Results

Quantitative yields of high purity products were observed for the entire series (Table 1). The output from the gas monitoring showed smooth uptake of hydrogen for all reactors and that the reaction is complete within 3 h. Figure 8 shows representative gas uptake curves obtained during the experiment, which are in good agreement with the predicted stoichiometric gas uptake of 1 mmol of hydrogen.

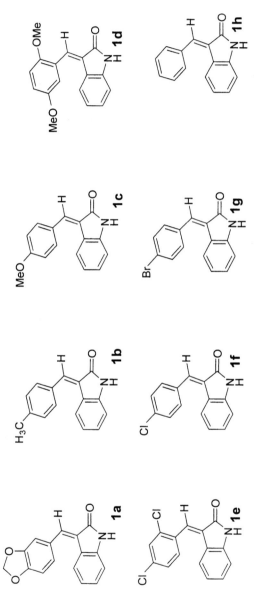

Figure 6- Materials used for parallel hydrogenation of 3-substituted indolin-2- ones.

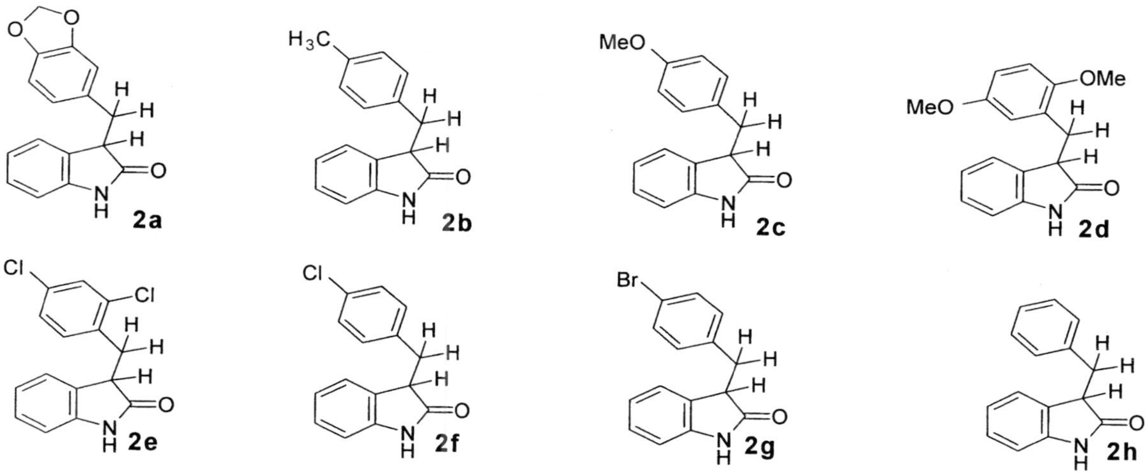

Figure 7- Hydrogenation products 2a-h.

Table 1-Hydrogenation results

Reduced Product	Yield(%)	HPLC Purity (%)
2a	98	90
2b	99	99
2c	99	94
2d	97	99
2e	99	92
2f	99	94
2g	99	94
2h	97	99

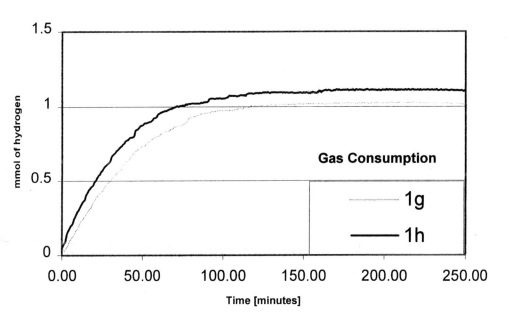

Figure 8- Representative plot of H₂ gas consumption vs. time for compounds 1g and 1h as described in the text.

Consistency of Activity Screen

Catalysts are sensitive to the small amounts of trace contamination, hence catalytic reactions are generally carried out in large reactors with small surface to volume ratios. Comparing the results between reactors for identical conditions tested the consistency of catalytic activity in the small reactors of the instrument.

Reaction Conditions

5 ml of a 0.58 M solution of nitrobenzene in methanol was added to a glass liner and inserted into each of the reactors. Approximately 18 mg of a commercial Pd/C catalyst[10] was added to each reactor. After purging, hydrogenation was carried out under 2 bar and 40° C. The gas consumption data was recorded and analyzed to determine the rate of hydrogen consumption per gram of catalyst.

Results

The consumption of hydrogen over time for all eight reactors is shown in Figure 9. The slope of this gas consumption curve is directly proportional to the rate of the reaction per unit catalyst. The rate was determined to be 0.01916 mol/l*s*g. The standard deviation between reactors was found to 3.6% for a single run. The reproducibility from run to run was found to be excellent, with average of two runs conducted by different operators showing a standard deviation of 4.7%. The final gas consumption showed a standard deviation of 6.0%, with individual values falling within 15% of the theoretical value, which is consistent with the experiences of other users of the instrument.

Screening for Chemoselectivity

Many of the small organic molecules used as pharmaceuticals or pharmaceutical intermediates have multiple bonds that can be hydrogenated. Chemically selective hydrogenation of these molecules is required for the desired transformation. Various candidates must be screened to obtain the catalyst with the appropriate selectivity. A simple molecule with no known pharmacological activity was chosen as a test case for the instrument.

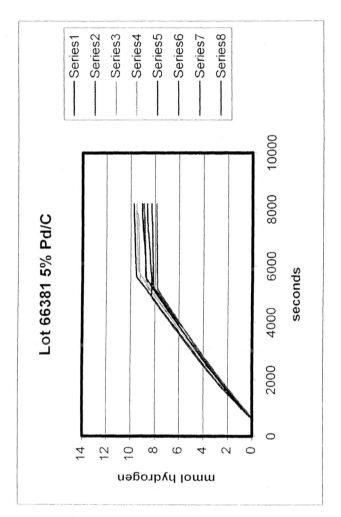

Figure 9- H₂ gas consumption vs. time for all reactors running simultaneously under identical conditions. The standard deviation in the rate of reaction is 3.6%.

Reaction Conditions

Trans-2-hexenylacetate has both a carbon-carbon double bond and ketone functional groups. A solution was prepared using methanol as a solvent. A variety of catalysts were screened: 3 types of Pd/C, Pearlmann's catalyst (also a form of Pd/C), two types of Pt/C, Pd/CaCO3, and Pd/Al2O3. Hydrogenations were carried out at 7 bar and 50° C. Gas consumption data were recorded until all the reactions were complete.

Results

The gas consumption as a function of time is shown for all reactors in Figure 10. All Pd/C catalysts showed identical uptake, less that the stoichiometric amount . In contrast other catalysts showed complete hydrogenation. Interesting the metal alone does not determine the degree of hydrogenation. The Pd/CaCO$_3$ and Pd/Al$_2$O$_3$ catalysts were the fastest and slowest catalysts which completely hydrogenated the substrate. The two Pt/C catalysts' rate was intermediate between the two oxide supported catalysts and also completed the hydrogenation.

Conclusion

A parallel pressure reactor system for catalyst screening was presented. Special consideration to certain aspects of the design was required to make a compact instrument for these types of measurements. Everyday concerns such as maintenance and safety were a major factor for specific design elements.
A variety of chemical applications of the instrument were presented. The utility of such a system for synthesizing various pharmaceutically relevant molecules and screening for activity and chemical selectivity was demonstrated.

References

[1] Rylander, P. " Hydrogenation Method," p. 29, Academic Press, San Diego, CA., **1985**
[2] For example: Johnson Matthey Catalytic Reaction Guide.
[3] Endeavor is a registered trademark of Symyx Technologies, Inc. The Endeavor and use of the Endeavor may covered by the following patents and patent applications: US 5,985,356; US 6,004,617; US 6,030,917; WO 00/09255;

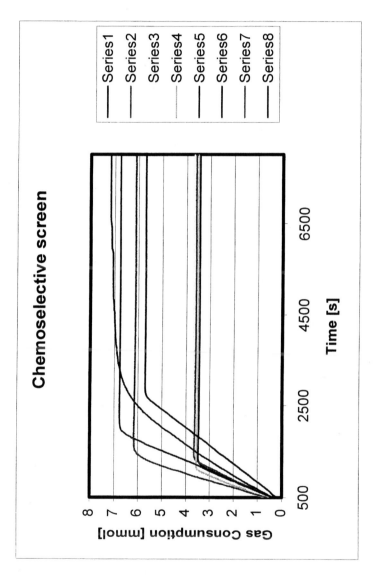

Figure 10- H₂ gas consumption vs. time for different catalysts used under otherwise identical reaction conditions. The catalysts show varying rates and different chemical selectivity's as discussed in the text.

US 09/177,170; US 09/417,125; US 09/211,982; US 09/239,223. Other patents pending.

[4] The difference between solution temperature and reactor body temperature is approximately 5° C. Each unit is individually calibrated, the average corrected temperature value is typically within 2° C of the actual solution temperature at equilibrium. Use of a heat transfer fluid between the glass liner and reactor body facilitates rapid temperature equilibration.

[5] Y. Sun; J. Wang; C. LeBlond, R.N. Landau; and D.G. Blackmond, *J. Catal.* **1996**, 161, 759.

[6] Temperatures in excess of 200° C are readily obtainable, however the perfluoroelastomer seals used degrade rapidly at temperatures above 200° C. Since sealing with high temperature compatible materials is substantially more difficult the operation was limited to 200° C.

[7] Sun, L.; Tran, N.; Tang, F.; App, H.; Hirth, P.; McMahon, G.; Tang, C. *J. Med. Chem.* **1998**, 41, 2588

[8] "Parallel Product Precipitation on the Quest 210: Synthesis and Purification of 3-Substituted-Indolin-2-ones Using New μFrit Reaction Vessels" *Argonaut Technologies Synthesis and Purification Letter # 103416.* Quest is a registered trademark of Argonaut Technologies, Inc.

[9] Reuvers, J.T.A.; DeGroot, A, *J. Org. Chem.* **1984**, 1110

[10] Sample courtesy of Johnson Matthey, Lot #66381.

Chapter 6

Conformational Analysis of Peptide Libraries

Saul G. Jacchieri

Fundação Antônio Prudente, Centro de Pesquisas, Rua Prof. Antônio Prudente 211, São Paulo, SP 01509–090, Brasil

Two independent, although related, approaches to the prediction of structural diversities of peptide libraries are shown. Extensive molecular and statistical mechanics calculations involving all components of tetrapeptide libraries are described. These results are compared to a knowledge-based conformational analysis of a library of protein structures containing a low sequence identity between any pair of proteins. An agreement between the results obtained with these two methodologies indicates a predominance of short-range and intra chain interactions in high structural propensity peptide fragments. The knowledge-based approach was applied to the complete set of 20^2 dipeptide and 20^3 tripeptide chain fragments enabling a detailed analysis of the structural diversities of these libraries. The application of these methods to the engineering of structural features in peptide libraries is discussed.

In the wide variety of biological functions due to polypeptide chains various intervening circumstances are self-directed. Proteins can fold spontaneously from a denatured state and carry out highly specific functions whereas peptides, by making specific interactions with proteins and other bio polymers, act as toxins, hormones, antibiotics, neuropeptides, etc...There is evidence, however, that proteins may have evolved from combinatorial libraries synthesized via a non coded (*1*), although to a certain degree (*2*) self sequencing, thermal polymerization of amino acids and that during the course of evolution most ancient proteins were excluded from biological systems. Considering an estimate of $3 * 10^5$ proteins in the human proteoses and of 20^{60} polypeptide chains of average length, it seems clear that proteins underwent a major screening, to use a term familiar to combinatorial chemists, of physical, chemical and biological properties.

Procedures, which do not necessarily follow the same path that has been attributed to the natural evolution of proteins but also include combinatorial and screening stages, have been devised to engineer specific features in polypeptide chains.

It is a fundamental principle of structural biochemistry that function is related to structure. Therefore, by inducing structure we would be able to reproduce existing functions or to create new ones. Combinatorial approaches have been applied to different aspects of the problem of designing structure and structure-activity relationships based on library composition. A combinatorial library whose components have conformations clustered near a single and pre determined conformation might be sought. These calculations may also intend to design libraries having a maximum range of activities and structures. They may also be focused on a desired activity, although allowing a variety of structures. A somewhat different object is the interaction between the components of a library and targets that may represent various kinds of receptors or other polymer chains. Some of the methodologies devised to design combinatorial libraries in accordance with these guidelines are discussed below.

An example of combinatorial calculations that intend to design a single conformation is a *de novo* synthesis (*3*) of the zinc finger. By generating sequence after sequence and maintaining the zinc finger main chain conformation, a computer-based combinatorial library is generated. The internal energy is used as a screening function to select sequences that stabilize the zinc finger. The best hit obtained in this combinatorial calculation was synthesized and shown to fold as a zinc finger. In these combinatorial calculations the total number of peptide sequences is pushed way beyond the maximum number of substances achievable in combinatorial libraries. Search algorithms (*4,5*) that efficiently avoid extensive sets of sequences inconsistent with the desired conformation are systematically employed to decrease the combinatorial size of the problem.

Such calculations have demonstrated that, as already known (*6*) in the field of protein structure, different amino acid sequences, and consequently a different packing of side chains and interactions between them, may result in the same structure. The combinatorial distribution of sequences among structures has been subjected to a theoretical investigation (*7*) that showed how it might be possible to estimate the library composition and size consistent with the adoption of a unique conformation.

A representation of library components as points in descriptor space is used to maximize molecular diversity. Diversity is defined (*8*) as a function of the distances between these points. More distant points correspond to more diverse library components. The chemical compounds belonging to a library are repeatedly replaced (*8*) with the aid of a simulated annealing algorithm until a maximum diversity is achieved.

Descriptor space is also employed to evaluate similarity with previously chosen lead compounds. In this case the intent (*9*) is the selection of a set of molecular building blocks that would narrow the molecular diversity. An alternative to descriptor space is the use of a quantitative structure activity relationship equation to distinguish chains having a desired activity. Targeting of a library of pentapeptide chains to a bradykinin analogue (*10*) is an example.

Activity is in many important cases due to the intermolecular interactions that result in molecular recognition. A simplified geometrical representation of molecular surfaces has been devised (*11*) to simulate interactions between library components and receptors, which may have a restricted or a broad surface complementarity. Maximum molecular diversity corresponds to library components having an overall complementary to a broad diversity of receptors, whereas targeting to a known receptor enables the selection of an activity.

The methodology described in the following sections is concerned with how the composition of amino acids affects the structures adopted by the components of a peptide library. The importance of peptide library composition is evidenced by the fact that, independently of sequence, the amino acids composition determines (*12*) the structural family to which a protein belongs. In order to engineer the properties of a peptide library, we would like to be able to predict structural diversity based on amino acids composition.

Structural diversity, referred to throughout this Chapter, is an important concept. A single polypeptide chain assumes a Boltzmann distribution of chain conformations near equilibrium. A peptide library also has a distribution of chain conformations that as a first approximation is the sum of the distributions available to the peptide chains that compose the library. In a structurally diverse peptide library the same main chain conformation may be adopted with a low probability by one of the library components and with a high probability by some other component, so that we may expect to find all main chain conformations. On the other hand, all peptide chains belonging to a low structural diversity peptide library have similar distributions of chain conformations.

The structural diversity problem is approached in two related ways. The property that unequivocally defines structural diversity is a complete set of sequence-main chain conformation probabilities. That is, the probability of occurrence of each main chain conformation adopted by each peptide chain belonging to the library. These probabilities may be calculated by making use of statistical mechanics techniques. Although this is a formidable task, reasonable approximations increase the feasibility of statistical mechanics calculations involving large numbers of sequences and conformations.

We may also explore the sequence and structural diversities contained in proteins. Presently, there are *circa* 15000 protein structures systematically stored (*13*) as numerical tables of coordinates. Many of the peptide chains belonging to a combinatorial library may be found as chain fragments in various proteins present in this vast collection of protein structures. In general, the shorter the chain fragments the higher its frequency in the database. Peptide fragments found in many structurally distinct proteins usually have a range of different conformations and this variability may be employed in the conformational analysis of single peptide chains as well as in the evaluation of the structural diversity of the entire peptide library.

These sequence-main chain conformation probabilities may be compared to their equivalents obtained theoretically. A comparison between the two sets enables an evaluation of the balance between short and long range interactions in library components.

The following three sections contain a description, based on a previous publication (*14*), of the methods employed in the calculation of sequence-conformation probabilities with force field and statistical mechanics techniques. The remaining sections, where a knowledge-based approach to conformational analysis is described, are being presented for the first time.

Indexes of Sequences and Main Chain Conformations

To address each structure allowed to the components of a peptide library we need unambiguous indexes that point to and, at the same time, describe sequences and structures. Each point in curves and surfaces constructed with these indexes corresponds to a peptide chain in the library and to a main chain conformation. There is an analogy between this geometrical representation and the structural diversity of the library, what gives to such indexes some of the attribute (*15*) of descriptors.

In the presently shown methods, peptide chain structures are generated by combining main chain and side chain rotamers. As a consequence, each peptide chain structure is a sequence of rotamers and, as shown below, similar indexes refer to sequences and to conformations.

Let us consider that A amino acids are involved in the combinatorial synthesis of the peptide library A_L. A_L is constituted by A^L peptide sequences of length L and the peptide sequences belonging to A_L are denoted by $a_i a_j a_k...$, where a_i is the ith amino acid listed in a table similar to Table I. In Table I the twenty natural amino acids employed in the combinatorial synthesis of the 20_L peptide library are listed. For reasons described below, frequencies of amino acid occurrence in a database of protein structures (*16*) are also shown in Table I.

Table I. Amino Acids Used in Equation 1

i	a_i	f_i
1	W	4710
2	C	5446
3	M	7589
4	H	7630
5	Y	11812
6	F	13068
7	Q	12388
8	N	14790
9	P	14934
10	R	16489
11	D	19207
12	I	19488
13	K	19846
14	T	19277
15	E	21554
16	V	23354
17	S	19443
18	G	24981
19	A	27148
20	L	27908

Natural amino acids listed in the order in which they are used to generate peptide sequences with Equation 1. i : index in the left-hand side of Equations 1; a_i : amino acid in the one letter code, f_i : frequency in a database (S_p) containing 914 protein chains with a low sequence identity between any two sequences.

The numeral L_S ($L_S=1,2,...,A^L$) indexes the components of A_L, and knowing L_S we may unambiguously generate the sequence $a_i a_j a_k$...by making use of Table I (or, if A < 20, of a shorter table) and the equations

$$i = int(\frac{LS-1}{A^{L-1}})+1$$

$$j = int(\frac{LS-1}{A^{L-2}}) - int(\frac{LS-1}{A^{L-1}}) * A + 1$$

$$(1)$$

$$k = int(\frac{LS-1}{A^{L-3}}) - int(\frac{LS-1}{A^{L-2}}) * A + 1 \ \cdots$$

Similarly, each main chain conformation is a combination of R main chain rotamers, seven of which are listed in Table II. The main chain conformations $c_i c_j c_k$... are indexed by the numeral L_C ($L_C=1,2,...,R^L$), and the main chain conformation corresponding to L_C is generated with the rotamers listed in (or in a table similar to) Table II by the Equations

$$i = int(\frac{LC-1}{R^{L-1}})+1$$

$$j = int(\frac{LC-1}{R^{L-2}}) - int(\frac{LC-1}{R^{L-1}}) * R + 1$$

$$(2)$$

$$k = int(\frac{LC-1}{R^{L-3}}) - int(\frac{LC-1}{R^{L-2}}) * R + 1 \ \cdots$$

Equation 2 is correctly used even if some main chain rotamers are not allowed to all amino acids. As discussed below, the ζ and ω main chain rotamers are allowed to proline residues, whereas the α and β rotamers are not. The only difference in this case is that main chain conformation probabilities containing disallowed rotamers are set to zero.

Table II. Main Chain Rotamers Used in Equation 2.

i	c_i	ϕ_i	Ψ_i
1	α	-57.0	-47.0
2	β	-139.0	-135.0
3	γ	-60.0	-30.0
4	δ	-90.0	0.0
5	ε	70.0	-60.0
6	ζ	-75.0	158.0
7	ω	-75.0	149.0

Main chain rotamers (c_i) listed in the order in which they are used to generate chain conformations with Equation 2. ϕ and ψ are main chain dihedral angles.

As indicated in Figures 3, 4 and 7, every sequence in the interval $i \leq L_S \leq i + A^{L-1} -1$ ($i=j^* A^{L-1} + 1$, $j=0,1,2,, A-1$) begins with the jth amino acid in (or in a table similar to) Table I. A similar periodicity is observed in the A^{L-2} sub intervals in which the above-described intervals are divided, and so on. The L_C axis is similarly divided in R intervals. This periodicity facilitates the interpretation of plots containing the L_S and L_C axis.

We may also want to include in the notation of a peptide library the A amino acids in its composition. In that case, as shown in Figures 3-6, A_L may be conveniently replaced by $\{a_i, a_j, a_k, a_l, \ldots\}_L$. This choice makes Table I unnecessary.

The correspondence between sequences main chain conformations and the indexes L_S and L_C are exemplified in Figure 1 for a tetrapeptide library.

Calculation of Sequence-Main Chain Conformation Probabilities

In Figure 1 internal energies and chain conformations belonging to the peptide library $\{L,K,E,I,V\}_4$ are shown alongside with the corresponding probabilities.

P_{SC} is the probability of adoption of conformation L_C by peptide chain L_S. A broad perspective of the structural diversity exhibited by the peptide library A_L is shown in $P_{SC} \times L_S \times L_C$ surfaces.

To calculate these probabilities we need a partition function for each peptide belonging to A_L

114

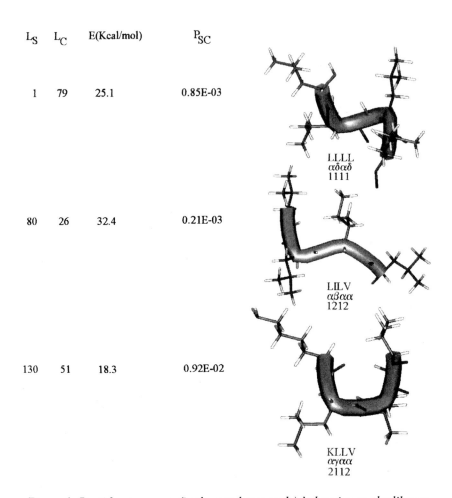

L_S	L_C	E(Kcal/mol)	P_{SC}
1	79	25.1	0.85E-03
80	26	32.4	0.21E-03
130	51	18.3	0.92E-02

LLLL
αδαδ
1111

LILV
αβαα
1212

KLLV
αγαα
2112

Figure 1. Peptide sequences (in the one letter code) belonging to the library {L,K,E,I,V}₄ and main chain rotamer conformations (listed in Table II) corresponding to selected values of the L_S and L_C indexes. Force-field energies (17) and sequence-main chain conformation probabilities (P_{SC}) are also shown. Side chain conformations are shown for completeness. The numerals 1,2 and 3 indicate, respectively, gauche minus, trans and gauche plus rotamers of the side chain dihedral angle χ_1. Solid tubes represent main chain conformations and stick models represent side chain conformations.

$$Z_S = \sum_{L_C=1}^{R^L} \sum_{\sigma 1=1}^{3} \sum_{\sigma 2=1}^{3} \sum_{\sigma 3=1}^{3} \ldots \sum_{\sigma_L=1}^{3} \exp(E_{SC}^{\sigma 1 \sigma 2 \sigma 3 \ldots \sigma L} / RT) \quad (3)$$

This partition function comprehends all conformations of peptide L_S compatible with the chosen set of rotamers. σ_i is the side chain rotamer conformation of the ith amino acid residue in the peptide sequence L_S and $E_{SC}^{\sigma 1 \sigma 2 \ldots \sigma L}$ is the force field internal energy (17) corresponding to the L_C main chain conformation and the side chain conformations σ_1, σ_2, σ_3,...,σ_L. The number of terms in Z_S, considering that we have five main chain and three side chain rotamer conformations, is $5^L * 3^L$. For other choices of a grid of rotamers this number may become very large, restricting the application of these calculations to short peptide chains.

We also need the statistical weight z_{sc} corresponding to the main chain conformation L_C.

$$z_{SC} = \sum_{\sigma 1=1}^{3} \sum_{\sigma 2=1}^{3} \sum_{\sigma 3=1}^{3} \ldots \sum_{\sigma_L=1}^{3} \exp(E_{SC}^{\sigma 1 \sigma 2 \sigma 3 \ldots \sigma L} / RT) \quad (4)$$

z_{SC}, due to the summation in side chain rotamers, is a statistical weight of a main chain conformation embedded, as illustrated in Figure 2, in an average of main chain conformations. P_{SC}, that corresponds to the relation

$$P_{SC} = \frac{z_{SC}}{Z_S} \quad (5)$$

, is the probability of occurrence of a main chain conformation. This relation includes all side chain conformations.

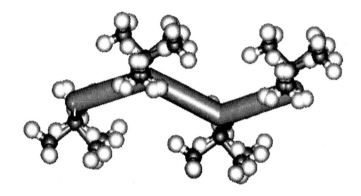

Figure 2. Averaging of Valine side chain conformations in tetra valine. A solid tube represents the β-strand main chain conformation. Superimposed ball and stick models represent the gauche minus, trans and gauche plus side chain rotamers of Valine.

A graphical representation of structural diversity

P_{SC} x L_S x L_C surfaces fully characterize the structural diversity of a peptide library and provide a broad view of structural changes accompanying sequence changes.

A high structural diversity is predicted by the structural propensities of the amino acids in the composition of the {A,M,F,G,P}$_4$ peptide library. Alanine and Methionine are α-helix forming amino acids, Phenylalanine, Glycine and Proline have a propensity to participate in reverse turns and Proline also has a propensity to participate in the poly proline I and II helices that resemble the collagen triple helix. This structural diversity is represented in the P_{SC} x L_S x L_C surface depicted in Figure 3.

The seven (R=7) main chain rotamers listed in Table II were employed in this conformational analysis. Figure 3 includes the structural motifs α-helix ($\alpha\alpha\alpha\alpha$ sequence of main chain rotamers), β-sheet ($\beta\beta\beta\beta$), reverse turn I ($\beta\delta\gamma\beta$), reverse turn III ($\beta\gamma\gamma\beta$) ,3$_{10}$-helix ($\gamma\gamma\gamma\gamma$), poly proline helix I ($\zeta\zeta\zeta\zeta$), poly proline helix II ($\omega\omega\omega\omega$), various random coil conformations and all tetrapeptide chains formed by the amino acids A,M,F,G and P.

There is, in Figure 3, a total segregation between the α-helix and the poly proline helix regions because the ζ and ω main chain rotamers are allowed only to proline residues which are not allowed to adopt the α and β rotamers. In most cases, only a partial segregation between distinct regions of P_{SC} x L_S x L_C surfaces is observed, showing that peptide sequences forming one structural motif may not form the other, and vice versa. This feature of Figure 3 suggests that the selection of the amino acids found in L_S intervals corresponding to partially segregated regions of P_{SC} x L_S x L_C surfaces may be used to design structurally homogeneous peptide libraries.

Besides this comparison between different regions of the same P_{SC} x L_S x L_C surface, comparisons between different peptide libraries are also useful. In Figure 4, where the {L,K,E,I,V}$_4$ and {A,L,V,M,I}$_4$ libraries are compared the L_C axis is the same for both libraries, although the L_S axis is not, because the same sets of main chain rotamers were employed in the conformational analysis of both libraries. As a consequence, the L_C cross sections depicted in Figure 4 show L_S intervals in which a given structural motif is selectively formed by one peptide library, and not by the other.

Combinatorial Libraries of Protein Fragments

Every protein fragment of length L has a sequentially homologous equivalent among the 20^L components of peptide library 20$_L$. A comparison (*14*) between the structural diversities of computer-based peptide libraries and

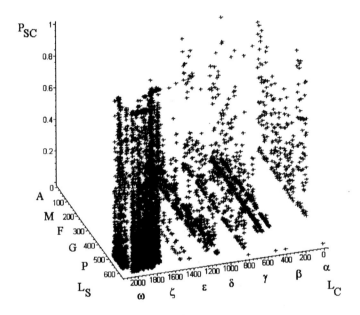

Figure 3. P_{SC} x L_S x L_C surface representing the structural diversity of the {A,M,F,G,P}$_4$ library. L_S intervals containing 5^3 tetrapeptide sequences beginning with the same amino acid are indicated by the corresponding one letter code amino acid. L_C intervals containing 7^3 main chain conformations beginning with the same main chain rotamer (as listed in Table II) are similarly indicated

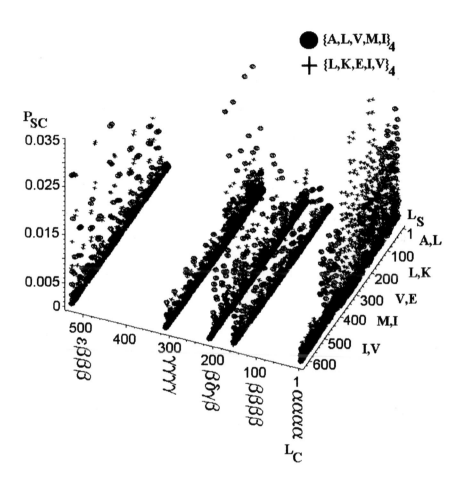

Figure 4. Superposition of P_{SC} x L_S x L_C surfaces belonging to the $\{L,K,E,I,V\}_4$ and $\{A,L,V,M,I\}_4$ libraries. L_S intervals containing 5^3 tetrapeptide sequences beginning with the same amino acid are indicated by the corresponding one letter code amino acid. Greek letters denote main chain rotamer conformations (listed in Table II).

of combinatorial libraries of protein fragments has shown that there is a coincidence of structural diversities, depending on factors discussed below.

This important, although limited, equivalence opens the possibility of using structural data extensively accumulated for proteins in the design of peptide libraries. For that purpose, we make use of the following definitions and formulae. S_p is a database of 914 protein structures with less that 25% sequence identity (16) between any two proteins. M_p is a randomized database of protein structures. M_p is obtained by shuffling each protein sequence in S_p, with the additional condition that the amino acid composition should be maintained, i.e., each protein sequence in M_p contains the same number of Alanine, Leucine, etc. amino acid residues found in its corresponding protein sequence in S_p. F_S is the frequency of peptide fragment L_S in S_p.

A comparison between results obtained with S_p and M_p enables a distinction between effects that may be ascribed to a random distribution of amino acids in proteins and effects due to the intrinsic properties of amino acids and to their intra and inter chain interactions.

A number of comparisons may be established between A_L (defined in the previous sections) and S_p and between A_L and M_p. One of the simplest quantities connecting sequences and structures is $F_S\chi$, the frequency of peptide fragment L_S when adopting conformation χ. χ is any of the structural motifs (α-helix, β-sheet, 3_{10}-helix, reverse turn I, reverse turn III, γ-turn and poly proline helices I and II) that may be generated with the main chain rotamers listed in Table II.

The ratio

$$PS\chi = \frac{FS\chi}{FS} \qquad (6)$$

is the evaluated probability of adoption of conformation χ by peptide fragment L_S in the S_p database of protein structures.

The equivalence between Equations 5 and 6 is investigated by actually calculating and comparing P_{SC} and $P_{s\chi}$. $P_{s\chi}$ is data derived from experimental data and includes the influence of short and long range intra and inter chain interactions, whereas P_{SC} is the result of statistical mechanics calculations that take into account only intra peptide chain interactions. The two quantities are, therefore, totally independent. A coincidence between P_{SC} and $P_{s\chi}$ indicates a predominance of interactions within the peptide fragment domain that may be used to create structure inducing templates employed in the design of structurally homogeneous peptide libraries.

Comparison Between Theoretical and Knowledge-Based Probabilities

For each component of a peptide library, we want to compare the probability $P_{s\chi}$ of occurrence of the structural motif χ in the library of protein structures S_P to the theoretical probability P_{SC} of the main chain conformation L_C, given that L_C and χ are structurally homologous. With the methods described in this chapter, it is possible to make detailed comparisons between the structural diversities of computer-based peptide libraries and the corresponding libraries of protein fragments.

There are evidences in favour and against (see, for instance, Befriends *et al* (*18*) and Minor and Kim (*19*) the conservation, regardless of chain environment, of structural propensities of peptide chains and homologous protein chain fragments. In the present calculations the dependency on the chain environment is part of $P_{S\chi}$, whereas the P_{SC} represent isolated chain fragments. An agreement between the two sets indicates that the interactions included in the force field calculations of the energy term in Equations 3 and 4 are predominantly important in the stabilization of the actual structural motif.

This comparison is extended to the whole range of amino acid composition available to the peptide library and to the library of protein fragments. The quantity

$$d_S = (P_{SC} - P_{S\chi})^2 \qquad (7)$$

is a measure, along the L_S axis, of the difference between both probability sets. d_s values approaching zero indicate that in the corresponding region of the L_S axis there is a similarity between conformational probabilities predicted by the described theoretical calculations and those determined in the database S_P. The average value of d_s (denoted by $<d>$) is an estimate of the overall agreement between P_{SC} and $P_{S\chi}$.

In Figure 5 this comparison is established for the α-helix main chain conformation and the peptide libraries $\{A,M,F,G,P\}_4$, $\{E,D,K,R,H\}_4$ and $\{A,L,V,M,I\}_4$. The $<d>$ averages are, respectively 0.09, 0.1 and 0.2.

Some of the strongest interactions that influence the calculation of P_{SC} in Figure 5 are the conformational constraints (*20*) of Proline, Phenylalanine, Valine and Isoleucine, which in most cases contribute to destabilize the α-helix, and electrostatic interactions between the ionized side chains of Glutamic Acid, Aspartic Acid, Lysine, Arginine and Histidine, which may be both stabilizing and de stabilizing.

The partial agreement between theoretical and knowledge-based probabilities seen in Figure 5 indicate that the electrostatic interactions and spatial overlaps built in most force fields may be employed to predict the α-helix content of a peptide library.

Figure 6, where a comparison between β-sheet probabilities is shown, leads to similar conclusions. Some of the most important interactions

122

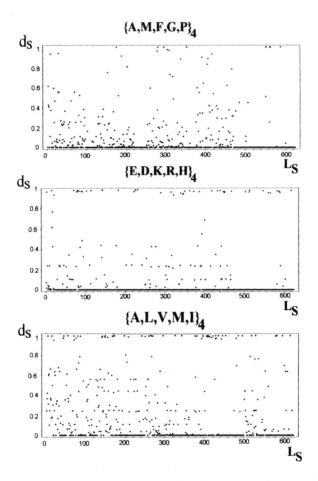

Figure5. Comparison, corresponding to the α-helix main chain conformation and to the peptide libraries {A,M,F,G,P}$_4$,{E,D,K,R,H}$_4$ and {A,L,V,M,I}$_4$, between $P_{S\chi}$ probabilities, calculated with Equation 6 in a database S_P of protein structures having a low sequence identity between any pair of proteins, and P_{SC} probabilities calculated with the aid of Equation 5. ds is defined in Equation 7 and the average values of ds are, respectively, 0.09, 0.1 and 0.2 .

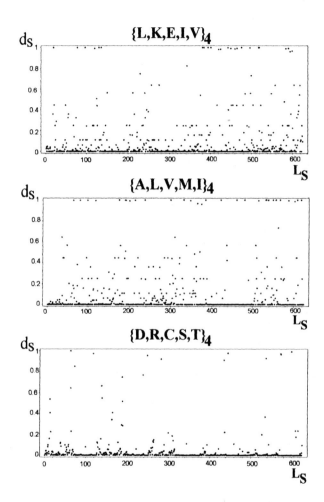

Figure 6. Comparison, corresponding to the β-sheet main chain conformation and to the peptide libraries {L,K,E,I,V}₄, {A,L,V,M,I}₄ and {D,R,C,S,T}₄, between $P_{S\chi}$ probabilities, calculated with Equation 6 in a database S_P of protein structures having a low sequence identity between any pair of proteins, and P_{SC} probabilities calculated with the aid of Equation 5. ds is defined in Equation 7 and the average values of ds are, respectively, 0.1, 0.1 and 0.04.

contributing to the calculation of P_{SC} in the peptide libraries $\{L,K,E,I,V\}_4$ ($<d>=0.1$), $\{A,L,V,M,I\}_4$ ($<d>=0.1$) and $\{D,R,C,S,T\}_4$ ($<d>=0.04$) are the conformational constraining of the beta branched (20) side chains of Isoleucine, Valine, and Threonine which have a tendency to stabilize the β-sheet and the electrostatic interactions between the ionized side chains of Glutamic Acid, Aspartic Acid, Lysine and Arginine. The plots in Figure 6 also indicate the possibility of predicting the content of β-strands in peptide libraries containing these amino acids.

Knowledge-Based Conformational Analysis

Besides the use of the above shown $P_{S\chi}$ probabilities of occurrence of structural motifs as a control of the reliability of theoretical calculations, $P_{S\chi}$ values alone may be employed in the conformational analysis of peptide libraries. The main advantages of the $P_{S\chi}$ knowledge-based probabilities are that they are entirely based on experimental data and $P_{S\chi}$ is evaluated by taking many protein chains into account, so that the influence of a particular chain environment may be averaged out if a sufficiently large number of non structurally homologous proteins are used to build the database S_P.

A comparison involving dipeptide and tripeptide libraries, the S_P and M_P databases and the α-helix and β-sheet structural motifs lead to various conclusions about a structural diversity and structural propensity.

The solid lines, which represent calculations with the M_P database, show in Figure 7 that if peptide fragments were randomly distributed in protein chains all fragments would adopt the same structural motif with similar probabilities.

The dots, which represent (non-random) distributions of peptide fragments in proteins, spread toward high and low $P_{S\chi}$ values. Their occurrence above and below the solid lines is an important deviation from random behaviour. Peptide fragments have high or low structural propensities because there are interactions (intra or inter chain or short or long range) that stabilize, or de stabilize, the corresponding structural motifs.

In regions in which random and non-random data overlap the distribution of peptide fragments among structural motifs is almost indistinguishable from random. Although in each particular protein structure peptide fragments are making specific interactions, the overall distribution is almost random in these regions.

It is seen in Figure 7 that whereas dots representing dipeptide fragments are almost uniformly scattered those representing tripeptide fragments are not. In the latter case, the existence of peaks in $P_{S\chi}$ x L_S plots is seen, indicating that tripeptide fragments belonging to well-defined intervals of the L_S axis have high structural propensities for the α-helix or the β-sheet.

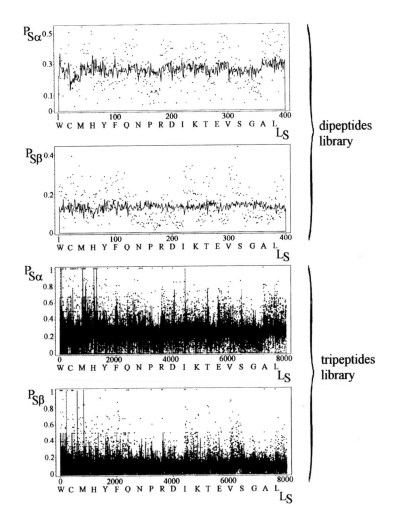

Figure 7. Probabilities, in the databases S_P (dots) and M_P (solid lines) described in the text, of occurrence (calculated in accordance with Equation 6) of the α-helix ($P_{S\alpha}$) and β-sheet ($P_{S\beta}$) structural motifs, plotted for 20^2 dipeptide and 20^3 tripeptide chains. L_S intervals where all 20 dipeptide and 20^2 tripeptide fragments begin with the same amino acid are indicated by the corresponding one letter symbol

In Figure 7, from left to right, the amino acids in the L_S axis obey the same increasing order of frequency in S_P listed in Table I. Hence, in the indicated L_S intervals, as well as in the whole L_S axis, the expected frequency of occurrence of tripeptide fragments is an increasing function of L_S.

It is clearly seen that the left-hand side of the tripeptide library plots is more crowded with P_{S_χ} maxima. Therefore, Figure 7 indicates that the proportion of amino acids participation in the peptide library is of critical importance and that if the proportions listed in Table I are not followed we may expect to obtain peptide libraries structurally dissimilar to the bio active peptide chains known to us.

Conclusion

It is in great part to the properties of amino acids and to interactions between them that proteins owe their structures. One of the main purposes of the combinatorial chemistry of peptide chains is to create tailor-made chains by following these principles.

In this Chapter, a study of the dependency on amino acid composition of the distribution of chain conformations among the components of a peptide library is presented. It is shown how a proper selection of amino acids causes an increase in the proportion of a certain structural motif within the library. A comparison is established between purely theoretical results and libraries, which we have named knowledge-based, of chain fragments belonging to protein chains found in the Protein Data Bank (13).

Although there are limited evidences (18,19) about the conservation of the conformations of protein fragments in sequentially identical peptide chains, a comparison between knowledge-based and wet lab libraries has not been done. We consider that this will be made possible with the availability of experimental information about the distribution of chain conformations in peptide libraries.

The purpose of these methods is to predict the amino acid composition necessary to engineer structural properties in peptide libraries. The next stage, from structure to function, has to be worked out case by case.

References

1. Rohlfing, D.L.; McAlhaney, W.W. *Biosystems* **1976**, 8,139.
2. Matsuno, K. *J Theor Biol* **1983**, 105, 185.
3. Dahiyat, B.I.; Mayo, S.L *Science* **1997**, 278, 82.

4. Goldstein, R.F. *Biophys J* **1994**, 66, 1335.
5. Gordon, D.B. ; Mayo, S.L. *Structure Fold Des* **1999**, 15, 1089.
6. Orengo, C.A.; Jones, D.T.; Thornton, J.A. *Nature*, **1994**, 372,631.
7. Zou J.; Saven, J.G. *J Mol Biol* **2000**, 296, 281.
8. Zheng, W.; Cho, S.J.; Waller, C.L.; Tropsha, A.. *J. Chem Inf Comput Sci* **1999**, 39, 738.
9. Cho, S.J.; Zheng, W.; Tropsha, A. *Pac Symp Biocomput* **1998**, 305.
10. Cho SJ, Zheng W, Tropsha A . *J Chem Inf Comput Sci* **1998** , 38, 259.
11. Wintner, E.A.; Moallemi, C.C. *J Med Chem* **2000**, 43, 1993.
12. Wang , Z.X. ; Yuan, Z. *Proteins* **2000**, 38, 165.
13. Berman,H.M.; Westbrook,J.; Feng, Z.; Gilliland, G.; Bhat, T.N.; Weissig, H.; Shindyalov, I.N.; Bourne, P.E. *Nucleic Acids Research* **2000**, 28, 235.
14. Jacchieri, S.G. *Molecular Diversity* **1998**, 4, 199.
15. Xue, L.; Bajorath, J., *Comb Chem High Throughput Screen* **2000**, 3, 363.
16. Hobohm, U.; Sander, C. *Protein Science* **1994**, 3 ,522.
17. Zimmerman, S.S.; Pottle, M.S.; Nemethy, G.; Scheraga, H.A., *Macromolecules* **1977**, 10, 1.
18. Behrends, H.W.; Folkers, G.; Beck-Sickinger, A.G. *Biopolymers* **1997**, 41, 213.
19. Minor, D.L. Jr.; Kim, P.S. *Nature* **1996**, 380, 730.
20. *Proteins*; Creighton, T.E., W.H. Freeman and Co.: New York, 1984; p 6.

Chapter 7

The Role of Informatics in Combinatorial Materials Discovery

Laurel A. Harmon, Steven G. Schlosser, and Alan J. Vayda

Nonlinear Dynamics, Inc., 123 North Ashley, Suite 120,
Ann Arbor, MI 48104

The authors argue for the importance of a broad definition of informatics in combinatorial materials discovery. The role of informatics encompasses three principal areas: flexible data management, data analysis and visualization, and predictive modeling. A challenge in data management is the variety and heterogeneity of information sources which must be utilized in the discovery process. The complexity and volume of combinatorial data demands new, more sophisticated analytical methods, although traditional statistical approaches are in wide use and exploratory data analysis tools are rapidly-evolving. The greatest informatics challenge is posed by the need for predictive materials performance modeling, which lies at the heart of an effective discovery process.

Introduction

A high-level view of the combinatorial discovery cycle is illustrated in Figure 1. *Experiment planning* drives the discovery cycle through the iterative design of individual experiments or series of experiments directed at a particular outcome. Synthesis and post-synthesis treatment of new materials is generalized as *library formation*. Materials characterization and performance testing is combined under the heading of *property measurement*. *Experiment interpretation* includes the reduction, visualization and analysis of experimental data. *Predic-*

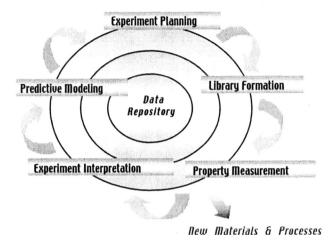

new materials & processes

Figure 1. The combinatorial materials discovery process is a cycle which relies on informatics at all stages.

tive modeling refers to the development of predictive relationships among experimental parameters, material characteristics, and desired properties using prior experimental and/or computational data. These relationships can then be used to develop the next stage of experiments. The role of informatics in this cycle is two-fold: 1) to provide the necessary tools for experiment planning, experiment interpretation and predictive modeling; and 2) to provide the underlying data management infrastructure with data capture and data delivery services. A full realization of combinatorial materials discovery requires an approach in which informatics components are integrated throughout the experimental process.

Issues in Combinatorial Materials Discovery

The application of combinatorial approaches for materials discovery presents a number of specific issues, many of which do not arise, or not to the same degree, in combinatorial drug discovery. These issues place new demands on informatics approaches. Among the special considerations of combinatorial materials discovery are the following:

- *Small-scale synthesis.* In order to result in meaningful discoveries, materials synthesized on a small scale must be "representative" of materials synthesized on a laboratory scale or in bulk. It is critical to establish criteria for being representative and methods to validate that these criteria are met.

- *High-throughput characterization.* Both the scale and throughput of a combinatorial approach pose challenges for the development of appropriate material characterization methods.
- *High-throughput assays.* Assays are required which are not only high-throughput but also sufficiently representative to truly predict the performance of scaled-up materials.
- *Process parameters.* Process parameters at all stages play an important role in material properties and performance.
- *Scale-up.* In order to have value, discovered materials must scale while maintaining their desired properties. Scale-up, then, is an essential consideration in the validation and optimization of methodologies as well as the assessment of potential material leads.

Two additional factors are of particular significance for the development of informatics components:

- *High-value experiments.* For many materials, the "expense" of combinatorial experiments can be high, in terms of time, reagent costs, instrumentation costs or other factors. This puts a premium on extracting as much value as possible from each "data point."
- *Limitations in first-principles modeling.* While computational chemistry has made major contributions to designing and analyzing combinatorial libraries of organic molecules, there are significant limitations in the ability to calculate relevant properties of many industrially-important classes of materials. As a result, more empirical approaches may be required to analyze and predict material properties and performance in the immediate future.

In addition to these technical challenges, a combinatorial approach to materials discovery requires methodological and cultural changes which are equally important for success. Examples of such changes include a transformation from a principal investigator research model to a team-based, multidisciplinary research model which incorporates the variety of expertise that must be brought to bear, including application-specific science and engineering knowledge, informatics experience, and laboratory automation capability. Experimental protocols must be made explicit and consistent so that data retain long-term value. Perhaps most importantly, combinatorial approaches enable researchers to explore beyond the boundaries of "common wisdom," but doing so requires challenges to fundamental assumptions and standard practice.

Key Informatics Challenges

Significant challenges are involved in each of the informatics elements of the discovery cycle illustrated in Figure 1.

Planning and Executing a Discovery Campaign

The task of experiment planning is to plan and execute a discovery program. The large numbers of composition and process variables which may be considered in a combinatorial approach mean that the discovery campaign must be a highly iterative, guided search. In the past, knowledgeable experimenters have guided the search by properly interpreting experiments *(1)*. More recently, automated experiment planning has been shown effective in some applications *(2)*. Either approach to planning requires methods to design experiments in high-dimensional spaces. Not only are there potentially many variables, but they may be quite different in kind. For example:

- composition variables may be treated as either continuous quantities or as block variables with fixed levels depending upon the goal of an experiment and the instrumentation available; and
- process variables such as temperature, pressure, magnetic field, may be either be smoothly varied or stepped, again governed by experimental objectives and by instrumentation.

A comprehensive experiment design capability must effectively handle a variety of parameter types within a single design. The need to navigate the discovery cycle many times in search of a particular material also requires the ability to aggregate or accumulate experiment designs over time in a systematic fashion.

An additional challenge to experiment planning stems from the downstream economic implications of differing material costs, energy requirements, and other factors influencing the commercial viability of a given material. Consideration *during the discovery process itself* of economic factors and the trade-offs between the benefits of material performance and the costs of material production and use may help to drive investigation more quickly toward commercially-relevant materials. This can be accomplished by incorporating these factors as constraints during experiment planning. Figure 2 illustrates the use of algorith-

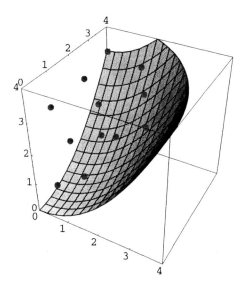

Figure 2. Physical, chemical, and economic factors can lead to complex constraints on experiment planning. This figure graphically illustrates optimal placement of experimental points in the space bounded by a curvilinear constraining surface in three dimensions.

mic methods to optimally cover arbitrarily-complex geometric regions arising from multiple simultaneous constraints.

Planning and executing a discovery campaign requires several key informatics elements:

- support for conceptualizing past and planned experiments in high-dimensional material and process space;
- the ability to view and retrieve data on the basis of single or multiple attributes and processes, aligned with the experimenter's mental models; and
- the integration of heterogeneous data sources.

Examples of technologies which can be brought to bear on experimental planning today include the following:

- sophisticated graphical user interfaces (GUI), which reduce the burden to the experimenter of specifying the scope and constraints of a given experiment;

- relational and object-oriented databases to consolidate and maintain complex relationships among experiment parameters and outcomes;
- constraint management approaches, to express and leverage expert domain knowledge;
- computational geometry, to design and evaluate plans in high-dimensional space; and
- knowledge-based planning approaches, to automate the analysis and resolution of complex trade-offs.

Developing Experiment Understanding

Combinatorial methodologies pose particular challenges to the interpretation of experiments and the development of new insights and understandings. These challenges stem from the numbers of variables involved as well as from the numbers of experiments themselves.

One important requirement is for automated data reduction and information extraction from experimental data, due to:

- the large number of experiments and data points, which prohibit typical manual and/or visual methods which are perfectly effective on an individual experiment basis;
- the need for results in a form suitable for storage in databases, either as quantitative data or as consistent qualitative or categorical descriptions; and
- the need for repeatable, reproducible measurements which can be compared over the lifetime of a combinatorial project and which can be interpreted and used by many members of a combinatorial team.

An additional motivation to automate data extraction and interpretation is the need for data which are suitable for empirical modeling studies, e.g. consistent data that are appropriate as input to statistical or pattern recognition algorithms (3). A complicating factor in achieving consistency is that instrumentation and experimental methods are likely to evolve, since this is not a "mature" field. At least a partial solution is to capture and retain as much information as possible about how data were generated and processed.

Automated data reduction does not, however, reduce the need for visualization tools. In fact, visualization tools are essential to review both raw and reduced data and to integrate information across different kinds of experiments. Visual data analysis plays an important role in the qualitative assessment of quantitative results and relationships, particularly the initial detection of potential trends and anomalies. Visual data analysis is also of value for exploratory analysis and building intuition.

There are significant limitations to visual data analysis:

- practical limits in the numbers of dimensions which can be usefully viewed and interpreted;
- a limited ability to assess multivariate relationships;
- the ability to derive only approximate, qualitative trends and correlations rather than precise, quantitative relationships; and
- variable interpretations of visual data between observers and over time.

In addition, the results of visual data analysis are not amenable to validation and archiving and the reliance on visual methods often impedes the development of an automated and integrated capability.

The reduction of data from a given type of instrument may or may not change in the transition from traditional to combinatorial methods. Instrumentation designed for combinatorial experiments may sacrifice signal-to-noise ratios, resolution or other factors in order to obtain higher throughput and parallelism. This, in turn, may lead to greater computational complexity in extracting reliable quantitative (or even qualitative) information from measurements.

The interpretation of combinatorial experiments requires several key informatics elements:

- algorithms to reduce data from individual instruments and experiments;
- automated quality control (QC) procedures including algorithms to extract and store measures of data quality;
- empirical data modeling and model evaluation methods; and
- visualization of N-variable interrelationships.

The technologies needed to support combinatorial experiment interpretation are relatively mature. Examples include the following:

- laboratory automation with automatic data capture, often supported by Laboratory Information Management Systems (LIMS);
- algorithmic multidimensional data analysis methods; and
- interactive multidimensional exploratory data analysis and visualization tools.

Leveraging Experiment Outcomes

The primary informatics challenge in leveraging combinatorial experiments is the need to generalize experiment results in a way that efficiently drives exploration and provides a basis for future experimental planning. Central to the combinatorial approach is the ability to develop screening techniques that focus attention on promising lead materials *(4,5)*. The development of effective screens depends upon:

- *Data characterization*: the application of descriptive analytical methods to capture general, quantitative representations of experimental results; and
- *Predictive modeling*: the development of quantitative models of the relationships between material characteristics, process parameters, and material performance.

Combinatorial materials discovery processes accelerate discovery partly through the parallelism of synthesis and testing activities. However, a more important gain results from the opportunity to systematically vary key variables such as composition and process within a common environment so that the *comparability* of the resulting data can direct the experimentation process. This can only be done if data characterization is of high quality, and also of sufficient breadth and representativeness that accurate predictions of material properties (performance measures) result.

In general, a data characterization process is designed to elucidate specific experiment outcomes as well as provide insight into factors that most influence those material properties of interest. Commonly, an experimenter can specify physical properties that are relevant and must be measured quantitatively or qualitatively in order to detect trends and correlations informative to experiment design. Beyond that standard practice, however, it is often possible to computer-process measurement data to obtain sufficient numbers of characterizing "features" having predictive value that experiment design can be focused on application-specific goals, and exhaustive search avoided. Such computer-processing is desirable in the context of combinatorial experimentation because it increases throughput, decreases reliance on subjective human judgements, and offers new information relevant to prediction.

Examples of such physical characterization processes are the use of scanning electron microscope images (SEM) to quantify surface characteristics such as roughness, and x-ray powder diffraction (XRD) to assess bulk properties such as crystallinity or phase. Both types of instrument are routinely used in materials discovery, but until recently most usage has relied on visual analysis of results. Greater leverage may be obtained from SEM data, for example, through the use of image processing to extract physical size, shape and surface properties; image normalization often enables the comparison of material properties across sets of related materials on the basis of absolute measurements rather than relative measurements. This approach has been applied successfully in the biological domain *(6)*. Greater leverage may be obtained from XRD data, for example, through the use of signal processing and pattern clustering to group (and possibly identify) materials prepared combinatorially when they have similar structural characteristics *(7)*. Both SEM and XRD characterizations that result from computer-processing offer rich sources of quantitative "micro-features" that can contribute to experiment selection and performance estimation. Figure 3 shows an example SEM image and the micro-features automatically extracted from it.

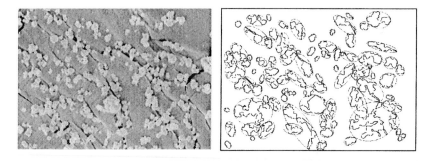

Figure 3. A SEM image can be automatically processed to extract quantifiable features.

On completion of an experiment and the collection of characterizing information and test assay results, it is possible to consider the construction of an empirically-derived predictive model that accounts for the relationship between inputs and outputs: composition and process variables and application-specific assay variables *(8)*. The value of such empirical data models is very high for they can be used to focus the discovery search as well as guide material and process optimization.

Response surface models are an example of empirical data models. Combinatorial experimentation is well-suited to response surface modeling since the systematic variation of composition and process variables is precisely that needed for response estimation (regression). In materials discovery, the ability to generate response predictions is often eroded by noisy laboratory measurements

138

(high variance), sparse data measurements (arising from a coarse-to-fine search strategy and the high-dimensional nature of the data), and nonlinear dependencies between factors and responses. To obtain the greatest leverage from the available data, it is important to employ methods that perform well in the face of these data characteristics. Typically, classical statistical tools such as multiple linear regression do not fare well in this situation. Fortunately, satisfactory alternative methods (neural networks, Gaussian processes, ridge regression and recursive partitioning) *(8,9,10)* are now available and do perform well in real-world applications. Typical aspects of these more powerful methods are portrayed through the use of simulated laboratory data in Figure 4. Here a nonlinear performance function is fit over a range of compositions (middle surface); the upper and lower bounding surfaces display the variance estimated at each composition point. As illustrated, modern interpolation methods simultaneously provide reliable estimates of mean and variance despite the presence of nonlinear relationships and correlated variables.

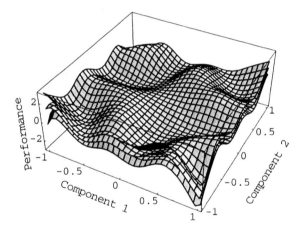

Figure 4. Response surfaces can be used to capture information about average quantities and distributions in complex spaces.

Additional leverage from combinatorial data may also be gained through using more powerful forms of inference than are typically used. Especially relevant to combinatorial experimentation are Bayesian approaches to statistical inference and data modeling that are not dependent upon restrictive parametric assumptions such as normality, equal variance and so forth. The recent introduction of Markov Chain Monte Carlo (MCMC) methods in regression and inference offer new powerful methods for the extraction of more information from each data set at the cost of additional computation *(11,12)*. Today, this is an attractive trade-off.

Flexible Data Management

The challenges which must be addressed by the data management system for combinatorial materials discovery include large volumes of data, a variety of data sources, a variety of data users, data organization, and change over time.

The most obvious data management challenge is the handling of a large volume of data. Manual operations that are acceptable under standard experimental conditions become onerous when the operations are multiplied by the 10s, 100s, or even 1000s of experiments that are conducted in parallel. Additionally, manual operations have the advantage that adjustments for minor (or major) discrepancies in the data can be easily detected and compensated for. The automation required to handle parallel experimentation depends on consistent data and operations. Consistency is also important for the analysis tools which utilize large volumes of data. On the other hand, a free-flowing discovery process will produce unanticipated views of the data and uncover previously unknown interrelationships which will require data and operations to evolve to take into account the knowledge gained over time.

In order to allow for changes in viewpoint over time, it is important to maintain complete data. Information that would typically be implicit or only exist in the mind or notebook of the experimenter must be represented explicitly in the database so that changes can be noted and taken into account. In a short-term series of experiments, it is easy to lose sight of the fact that only those parameters which have been explicitly recorded are available for future analysis and comparison. Of necessity, then, more information must be retained than those factors whose importance was anticipated, if new insights are to be gained over time. Completeness also makes it possible to perform analyses on data that were originally collected for other purposes. Change to the system and procedures over time will also require changes to the database and the tools which work with it. The data management system must accommodate updates without making older data unusable. This requires flexibility throughout the data management system in capabilities such as data import, data representation, and data access.

Combinatorial data require different organizational principles than traditional data. For example, in the traditional approach, a single material is synthesized, characterized, or assayed with each equipment "run." Under the combinatorial approach, an equipment "run" may involve many samples. In this case, it is crucial to track the location of each material sample in the run. Each sample will participate in many different runs on various pieces of equipment with different locations each time. Thus, the collection of all of the information regarding a specific material sample requires a complicated extraction process. The data integration problem is further complicated by the many possible data sources with their individual specifications and the many users with their individual requirements. Figure 5 shows examples of the variety of data sources and users which must be supported.

The solution to this problem is to adopt a sample-centric approach to data storage. When stored in the database, data from multi-sample runs are broken up into sample-sized units. Additionally, the sample-sized units arising from different runs will be linked together for each sample. Run level information is maintained in this approach but it is secondary to the sample level information. Each type of equipment run will produce results in its own output format. Interpretation, translation, and standardization of the output data will be performed upon import into the database.

Integration of data from a variety of sources is crucial to successful application of analysis tools. A combinatorial materials discovery system will have many types of equipment which require input and provide output in different formats. In order to build a model relating, for example, material synthesis parameters, material characterization results, and material performance measurements, it is necessary to assemble all of the data from the samples of interest and the equipment runs of interest and feed it to the analysis tool in a uniform manner. The sample-centric approach addresses this challenge during the data import process thus simplifying the problem of extracting data for analysis. Each analysis tool will have different requirements for its input data but they will pull their data from a uniform representation which is independent of equipment specifics or data format issues.

Another integration issue for combinatorial materials discovery is the need to integrate with resources outside of the combinatorial system or to incorporate information which originated from other sources. For example, historical data may be available which are relevant to classes of materials under investigation, but which have not been developed with a combinatorial approach in mind. While it may not be possible to incorporate these data in exactly the same fashion as the combinatorial data, they may be included as part of a background knowledge base which is accessible to algorithms and researchers. Similarly, special studies of material leads using non-combinatorial analytical instruments, or higher-volume synthesis methods need to be brought into the system. In some

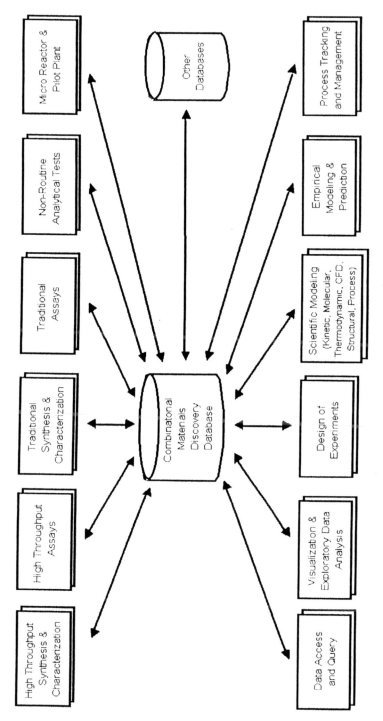

Figure 5. The variety of data sources and users requires a data management system which is capable of managing and integrating diverse data types.

organizations, this may also require integration or interface to Laboratory Information Management Systems (LIMS) or databases from other parts of the organization.

Integration and Implementation Issues

Emerging Integration and Infrastructure Technologies

The implementation of the infrastructure to support the informatics needs of combinatorial materials discovery is facilitated by taking advantage of the latest advances in information technology. The most important areas include databases, communications, and development platforms.

While a traditional relational database can be used to support the rapidly evolving needs of combinatorial materials discovery through careful design, object-oriented databases are better suited to handling the diversity of data types and the inevitable database changes. As relational database vendors build more object-oriented properties into their relational databases, these new object-relational databases provide some of the benefits of object-oriented databases without giving up the benefits of the relational approach *(13,14,15)*.

The adoption of internet communication protocols by all computer vendors and the opening up of many previously closed systems by software vendors has greatly enhanced the capabilities that can be built into new informatics systems. Products from various vendors can be integrated easily, resources can be shared through distributed systems, and multiple computer hardware and operating systems can be supported transparently.

At the center of this revolution are web-based systems which utilize the ultimate open, cross-platform, distributed development platform. Most software vendors are opening up their systems and offering web capabilities which simplify the development and integration process.

Communications advances have also had an impact on the human part of the combinatorial materials discovery system. Networking and collaboration tools such as email, instant messaging, groupware, and video conferencing facilitate interaction between the core team members, other resources within the same company, resources at other company locations, and, through the use of virtual private networks (VPN), resources at other organizations.

Implementation Strategies

A conventional approach to the software implementation of an informatics system tailored to combinatorial materials discovery is to determine functional requirements based on an in-depth understanding of the application, design the software architecture to meet these requirements, and then implement that architecture from beginning to end, module-by-module, testing and validating each module in turn. This methodology, often called the waterfall model, is well-suited to applications for which requirements are well-defined. However, combinatorial discovery systems are newly in use and their software requirements not yet well-understood. Consequently, the standard approach to software development is not recommended, and an alternative approach necessary.

In contrast to the waterfall model, the spiral model of software development *(16)* is suited to situations in which requirement understanding is initially incomplete but improves as experience is gained with an implementation. A spiral approach to discovery systems is effective since it provides an opportunity, first, to prototype a simplified end-to-end system, then to evaluate that prototype and, subsequently, deepen the implementation according to new understandings and requirements. This cycle may be repeated to whatever degree is necessary, thus giving rise to the concept of a spiral process: one which starts small and progressively widens to encompass ultimately the entire application domain.

The diagram in Figure 6 illustrates the main features of a spiral development process:
- analysis,
- implementation,
- validation,
- refinement, and
- documentation and future plans.

In addition to these phases of development, it is typical to undertake at least three turns on the spiral: a proof-of-concept stage (POC), a stage that provides an initial operating capability (IOC), and then subsequently a stage that establishes a final production-level operating capability (FOC). Cases of exceptional complexity may, in fact, call for even more steps to achieve a final system, fully adequate for combinatorial materials discovery.

144

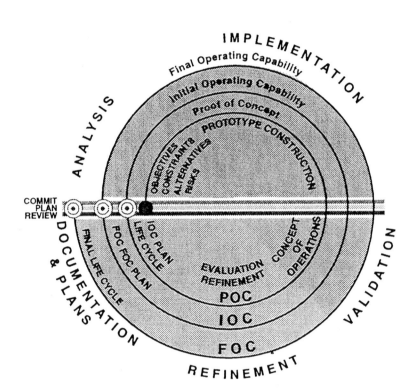

Figure 6. The spiral development process is based on successive refinements of an end-to-end system, supporting an evolutionary approach to technology development.

References

1. Myers, R.H.; Montgomery, D.C. *Response surface methodology: process and product optimization;* John Wiley and Sons: New York, 1995.
2. Wolf, D.; Buyevskaya, O.V.; Baerns, M. *An evolutionary approach in the combinatorial selection and optimization of catalytic materials*; Applied Catalysis A: General, **2000,** *200,* pp. 63-77.
3. Ripley, B.D. *Pattern Recognition and Neural Networks*; Cambridge University Press: Cambridge, UK, 1996.
4. Brocchini, S.; James, K.; Tangpasuthadol, V.; and Kohn J. *Structure-property correlations in a combinatorial library of degradable biomaterials*, J. Biomed. Mater. Res. **1998,** *42,* 66-75.
5. Brocchini, S.; James, K.; Tangpasuthadol, V.; and Kohn J., *A combinatorial approach for polymer design*; J. Amer. Chem. Soc. **1997,** *119(19),* 4553-4554.
6. Vogt, R.C., Trenkle, J.M. and Harmon, L.A. *Mosaic construction, processing, and review of very large electron micrograph composites*; International Symposium on Optical Science, Engineering, and Instrumentation, Proceedings SPIE 2847, Denver, CO, 1996, pp. 2-15.
7. Bem, D.S., et.al. *Approaches to Combinatorial Discovery of Materials via Hydrothermal Synthesis*; Combi 2000 Conference: Combinatorial Approaches for New Materials Discovery, January 23-25, 2000, San Diego, CA; The Knowledge Foundation: Boston, MA.
8. Neal, R.M. *Bayesian Learning for Neural Networks*; Springer Lecture Notes in Statistics; Springer-Verlag: New York, 1996.
9. MacKay, D.J.C. Ph.D. thesis, California Institute of Technology, Pasadena, CA, 1991.
10. Williams, C.K.I.; Rasmussen, C.E. In *Advances in Neural Information Processing 8;* Touretsky, D.S.; Moser, M.C.; and Hasselmo, M.E. Eds.; MIT Press: Boston, MA, 1996.
11. Gilks, W.R.; Richardson, S.; Spiegelhalter, D.J. *Markov Chain Monte Carlo in Practice;* Chapman & Hall: London, 1996.
12. Falcioni, M.; Deem, M.W. *Library Design in Combinatorial Chemistry by Monte Carlo Methods;* Phys. Rev. E, **2000,** *61,* 5948-5952.
13. Cattell, R.G.G. *Object Data Management*; Addison-Wesley: Reading, MA, 1994.
14. Chaudhri, A.B.; Loomis, M.E.S. *Object Databases in Practise*; Prentice Hall PTR: Upper Saddle River, NJ, 1998.
15. Stonebraker M.; Brown, P. *Object-Relational DBMSs*; Morgan Kaufmann: San Francisco, CA, 1999.
16. Boehm, B.W. *A Spiral Model of Software Development and Enhancement;* Computer, May, 1988, pp 61-72.

Chapter 8

Developing Combinatorial Support for High-Throughput Experimentation Applied to Heterogeneous Catalysis

D. Demuth[1], K.-E. Finger[1], J.-R. Hill[2], S. M. Levine[3],
G. Löwenhauser[2], J. M. Newsam[4,*], W. Strehlau[1], J. Tucker[5],
and U. Vietze[1]

[1]hte Aktiengesellschaft, Kurpfalzring 104, D-69123, Heidelberg-
Pfaffengrund, Germany
[2]Molecular Simulations GmbH, Inselkammerstrasse 1, D-82008
Unterhaching, Germany
[3]Molecular Simulations Inc., 9685 Scranton Road, San
Diego, CA 92121-3752
[4]hte North America, 6540 Lusk Boulevard, Suite C276,
San Diego, CA 92121
[5]Molecular Simulations Ltd., 230/250 The Quorum, Barnwell Road,
Cambridge CBS 8RE, United Kingdom
[6]Corresponding author: john.newsam@hte-company.com

Computational aspects of high throughput experimentation
(HTE) in application to heterogeneous catalysis are
considered. Specifically, design issues associated with
implementing a suitable MatInformatics system, and
approaches both to the Design stage, in which sets of
experimental points to sample are selected, and to the Model
stage, in which accumulated data are interpreted in the context
of predictive models, are discussed. Citations of recent reports
are used to illustrate progress; some opportunities for future
development are also outlined.

Computation is today regarded almost universally as a critical component of high throughput experimentation (HTE), but there is, as yet, no complete solution to the various computational challenges that arise in applying HTE in routine practice. This represents a significant research and development opportunity. Accordingly, we highlight here some key needs and illustrate recent progress towards addressing these needs.

The focus of the following discussion is heterogeneous catalysis, one of the main emphases of our own current programs; very similar issues arise in applications of high throughput experimentation to many other materials science areas. We consider here the computational elements, but in so doing underscore that these are very much integral to the high throughput experimentation cycle, a key aspect of which is the integration of different technologies and different disciplines in focussed application. Several recent overviews (*1-7*) offer an entry point into the broader literature that describes developments and applications of the various experimental components.

High Throughput Experimentation - Combinatorial Methods in a Broader Context

Although the term is frequently misused, or perhaps abused, 'combinatorial chemistry' has a sensibly explicit definition: *Combinatorial chemistry is the production of libraries of compounds that represent permutations of a set of chemical or physical variables.* The chemical variables might be attributes of the *products* in the compound library, for example sets of possible R-group substituents at each of a set of defined positions on a molecular scaffold, or sets of discrete substituents or average compositions of substituents at a specific

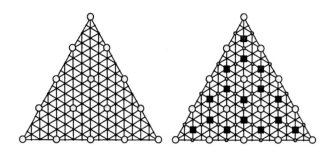

Figure 1. A true combinatorial library represents a grid search in the experimental variables space. In this schematic library with three variables, to increase the sampling density in the first design (left, circles), the intervening elements drawn as squares (right) cannot alone be selected; rather the expanded matrix, with all large and small circles is, for example, sampled in the strict combinatorial case.

cation site in an inorganic solid. To produce such libraries, though, requires that a rational or product-directed synthesis route is available. The chemical variables might, alternatively, be *precursor compositions or sets of reaction stages* that can be applied to an initial substrate or substrates; we are today almost always in this latter category when producing true combinatorial libraries of heterogeneous catalysts. We might also be sampling physical variables, such as the temperature or pressure used in a given synthesis procedure, or perhaps a stirring or deposition rate. The variables may be discrete, for example a set of possible alkyl substituents at a defined position in a cyclopentadiene ring, or continuous, for example average composition in a continuous solid solution or physical variables such as temperature, pressure, or stirring rate. By strict definition, a combinatorial library contains the products reflecting permutations of all of the sampled possibilities for one variable with all of the sampled possibilities for all of the other variables. This 'combinatorial constraint' defines a full grid-search in the variable space, generally an inefficient sampling approach (Figure 1) (*8, 9*). Hence, unless our synthesis method is intrinsically permutative, such as the pooled synthesis of molecular catalysts, or offers substantial efficiency gains if applied in a permutative fashion, we will usually elect not to produce true combinatorial libraries.

In current practice, though, the term 'combinatorial chemistry' is used rather loosely and, as a result, might have several intended connotations. The term is sometimes intended to describe compound library synthesis in general and, occasionally, to even include *high throughput screening – the rapid assessment of a specific property for each of a large number of samples.* Synthesis or screening in isolation is of limitedpractical interest; it is integration of synthesis and screening, together with other elements (Figure 2) that proves of most value, hence the preferred use of the term 'high throughput experimentation', particularly in chemicals and materials applications.

High Throughput Experimentation is the rapid completion of two or more experimental stages in a concerted and integrated fashion. High-throughput experimentation typically comprises four interconnected stages. Expressed as actions, the HTE cycle comprises "Design", "Make", "Test" and "Model" stages (Figure 2) and this cycle or spiral applies equally to the discovery and development of drugs, heterogeneous catalysts, or other materials. These prime stages are considered further below.

The HTE cycle would be essentially unworkable without a MatInformatics system to manage the data associated with all of the operations involved. Computation, though, impacts well beyond this 'operations management' role, in areas such as robot and instrument control and monitoring, project management and reporting, instrument design and simulation. The key role of computation is particularly obvious in the Design and Model stages.

150

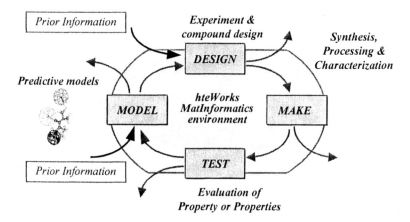

Figure 2. Schematic of the high throughput experimentation cycle, illustrating the coupling of Design and Model stages into the experimental stages of catalyst and catalyst library synthesis and characterization, and the assessment of catalytic activity and selectivity (copyright hte Aktiengesellschaft, reprinted from www.hte-company.de with permission).

The High Throughput Experimentation Workflow

A characteristic of the application of HTE in the materials arena is diversity – diversity in materials compositions, types, and methods for synthesis, processing, characterization and property assessment. Accommodating such diversity is a design requirement on a corresponding MatInformatics system. Even within heterogeneous catalysis, the ostensibly small subset of the materials field considered here, the details of specific HTE programs vary hugely. Synthesis and processing methods, characterization techniques and, particularly, reactor designs need to be tailored for each reaction and catalyst class.

In applications of HTE in heterogeneous catalysis, we distinguish between between two domains, Stage 1 and Stage 2 (6); the computational demands are generally different in these two stages. In Stage 1, where the target might be to discover a catalyst effective in promoting an unmet conversion need, such as the conversion of benzene to phenol in the gas phase, we will usually seek to produce very large libraries of materials, each element in a small amount, and then probe a property of each member that can be interpreted as an initial signature of catalytic utility. This signature might be a measure of total exothermicity (10-12), or the yield of specific product molecules, measured by, for example, mass spectrometry (13-15), infra-red (16) or resonance enhanced multiple photon absorption (REMPI) (17, 18). As the performance of a heterogeneous catalyst represents the overlap between (1) the attributes of the catalyst itself, (2) the reactor design, and (3) the operating conditions, however, Stage 1 testing conditions are frequently limited in practical value. In Stage 2 testing, on the other hand, we seek to evaluate catalysts under 'real conditions', that is, under similar conditions to those applied in conventional laboratory testing. This implies operating under suitable pressure and temperature, perhaps using real feeds or at least models that approximate the real feeds, practical catalyst morphologies, and with a detailed or full analysis of products. We gain efficiencies, relative to traditional experimentation, through miniaturization, parallelization, automation, robotics, simulation and MatInformatics.

Although, this Stage 1 – Stage 2 distinction is generally accepted (7), it is perhaps not broadly appreciated that the two Stages have near-complementary roles in catalyst discovery. The throughput in Stage 1 might be some 500-1,000 samples per day; that of Stage 2 perhaps some 50 samples per day, depending on the complexity of the chemistry. However, though the Stage 1 throughput is substantially higher, there are many cases where the more controlled, understood and informative regime of Stage 2 is preferred as the entry point in a catalyst discovery program.

In most heterogeneous catalysis HTE programs, be they commencing in Stage 1 or Stage 2, an initial goal is to achieve 'production mode' with the complete system (Figure 2); this requires that all components are functioning reliably, that bottlenecks between components have been resolved and that the target throughput is achieved reliably on a daily or weekly basis. In Stage 1, much of the pre-'production mode' effort can be invested in synthesis procedures; in Stage 2 it is more usually the reactor and product quantitation system - we typically require, for example, that the reactor system support 24 hours a day - 7 days a week operation. The MatInformatics system must, naturally, support these requirements and have the flexibility that any special needs of the specific case can be smoothly accommodated.

Once the system is operational in this 'production mode', the flow illustrated in Figure 2 then applies. The individual or team responsible for the given program first makes decisions as to the next series of experiments to perform – which catalysts to produce, how they will be processed pre-testing, how they will be characterized, how they will be tested. This "Design" step leverages various computational tools, such as factorial design and other design of experiment ("DOE") protocols, the evaluated results of past rounds of experiments, information already available from other sources, and the insights and intuition of the project team. It is rarely 'blind testing', even if the designs are typically engineered to include a suitable proportion of non-conventional ideas. The output of the Design step is an experimental Agenda, reduced first to a set of instructions for the robotic device(s) used for catalyst synthesis. The robotic device automates repetitive tasks so that, applied by the skilled practitioner, catalyst libraries can be produced reliably. This "Make" stage (Figure 2) also, particularly in Stage 2, includes analytical characterization, by, for example, powder X-ray diffraction applied to arrays of samples, and post-synthesis modification, such as hydrothermal conditioning, catalyst aging or deactivation. The catalyst testing profiles defined in the Design stage are typically applied in a parallel reactor system (the "Test" stage). The data relating to and produced by all of these operations are housed in the MatInformatics system, where they can be assessed by the project team. Various visualization tools are almost always used in this process, but data mining protocols, and other model development tools (see following) can also be valuable, particularly when more voluminous data sets have been accumulated. This "Model" stage is, then, coupled closely with the Design stage in application within the next iteration through the HTE cycle.

A key design requirement initially imposed on the MatInformatics system is that it support this workflow; the MatInformatics system also provides the environment into which can be integrated current and future approaches to the Design and Model stages.

MatInformatics in the HTE context

The term 'Informatics' refers to the science and technology of operations on information – this strict, but perhaps unhelpful, definition builds on definitions also of the terms datum, data, information and knowledge. Pragmatically, 'Informatics' loosely encompasses the methods and the technologies associated with processing, storage and retrieval, management, dissemination and interpretation of data. 'MatInformatics' is a subset that refers to informatics for the broad domains of chemicals and materials, outside of the realm of molecular organic entities (the latter being addressed specifically by 'cheminformatics').

MatInformatics, while still fragmented as a field, is developing swiftly. Its prominence reflects the burgeoning impact of information technologies in general. In a materials research context, the requirements of and interest in HTE are, though, fueling progress.

Design - Structural and Experimental

Issues associated with the application of design methods to heterogeneous catalysts, in the context of high throughput experimentation, have been considered recently (*19*). Three prime theme areas are – (1) design methods in general for heterogeneous catalysts, (2) 'library design' tools analogous to those used in high throughput experimentation in drug discovery, and (3) methods for selecting points to sample within complex multi-dimensional spaces. Literature on the first is reasonably extensive and has a significant history. Heterogeneous catalyst 'design' traditionally refers to the application of design principles to one or more of three elements: structure, function and synthesis. The interrelationship between these elements is usefully considered via a system's view (*19-22*). The *performance* of a catalytic system represents the combination of a set of properties. A specific *property* is governed by the geometrical *structure* of the total system, and of its components, at atomic, microscopic, mesoscopic and macroscopic length scales (catalytic activity is fundamentally an atomic-level phenomenon which is governed by electronic structure, but the ground state electron density distribution is uniquely defined by the geometrical structure). Structure is controlled by the conditions of *synthesis and processing.*

Heterogeneous Catalyst Design

Design implies balance and, for heterogeneous catalysts, factors that impact this balance include activity, productivity, selectivity (to the target molecular products, as well as regioselectivity and enantioselectivity), regenerability, durability and other mechanical attributes, ease of handling, separability, availability, and material and process economics. Even in development simply of the catalytic system itself, many issues are involved (see, e.g., (20, 23)). As a result, approaches to heterogeneous catalyst design have been diverse (24, 25), as illustrated by considerations of specific catalytic systems such as for olefin epoxidation (26), copper-based reduction (27), metal-support systems (28) and oxidation catalysis (29). Molecular-level design has been applied in optimization of surface structure (30, 31), molecular modifiers for enantioselective catalysts (32, 33), anchored molecular entities (33, 34), and design of active site environments in microporous crystals (35, 36).

Molecular simulation provides a useful atomic-level design framework, as today's molecular simulation methods allow the viability of particular structural arrangements to be gauged, and many of the properties associated with particular structural arrangements to be computed (37). For crystalline microporous solids, such as zeolites, detailed structural data are often available. Even if these do not refer to the exact system under study, such data provide a framework within which to apply various molecular simulation methods and a basis on which the reliability of such methods might be assessed. Design studies already published include development of virtual libraries of framework structures with desirable pore architectures (38), studies of zeolites with one-dimensional channels for hydrocracking (39), predictions of the geometrical effects of particular aluminum distributions (40), and simulations of non-framework cation positionings (41-43). Molecular simulation has been applied effectively also to other classes of heterogeneous catalysis, including hydrodesulfurization (44) and high temperature water gas shift (45), and, for example, to fundamental studies of zirconia (46), hematite (47), and the physi- and chemisorption of small molecules on metal surfaces (see, e.g., (30, 48, 49)). Reports of the early interlinking of atomic-level design and 'combinatorial' approaches have also begun to appear (50).

Creating and Screening Virtual Libraries of Inorganic Solids

To pursue a structure-based library design approach, a means of first generating virtual libraries of structures is needed. In considering methods then for developing virtual libraries of inorganic solids, akin to the methods developed for molecular organic entities, our preference is towards methods that are *systematic* and which yield *sensible* and, potentially also, *suitable* structures; that is, that the structures be viable in physicochemical terms, and that they manifest, or be likely to manifest, the properties of interest. For a structure to be viable it need represent a local minimum on the free energy hypersurface, but not necessarily the global minimum. Which minima on the free energy hypersurface will be accessible synthetically is not known *a priori* and the possibility of particular structures cannot be ruled out based solely on an ostensibly high internal energy value. The fourth design requirement, that the structures be *synthesizable*, cannot be imposed at this point, as the ability to develop rational synthesis of crystalline inorganic solids is today almost completely lacking.

Within the first three requirements above, virtual libraries of inorganic structures may, in select cases, be developed by *enumeration* or by *sampling* (descriptions of these approaches and citations of recent applications are provided in (*19*)). Based on each of the hypothetical structures produced by these methods, crystallographic data or analytical fingerprints such as powder X-ray diffraction patterns can be computed and then compared with suitable databases to identify which elements in the virtual library have already been observed in practice. The virtual library elements might also be screened computationally, based on their computed properties, as for the molecular case. One advantage of using Monte Carlo based methods for sampling structures in a defined possibility space (*19*) is that it is straighforward to include additional terms into the 'energy' or cost expression that is the basis for structure development. Thus, the degree of match with target analytical data (*38*) or the matching of a target set of properties such as pore dimensions (*38*) can be used to help polarize the structure search towards more attractive structural candidates. A virtual library screening in these cases is thus built intrinsically into the virtual library creation process.

A notable example of Monte Carlo based sampling, and perhaps a small step towards linking virtual library creation and synthesis, is the recent description of a route to the *de novo* prediction of inorganic structures through an automated assembly of secondary building units (AASBU) (*51*). This method samples the ways in which a defined structural motif or set of motifs, such as a pentameric cluster of apex-shared octahedra, can be interlinked with replicates in 3-dimensions so as to form periodic framework structures. In contrast to earlier Monte Carlo based sampling methods, that use discrete tetrahedral and/or octahedral units and defined unit cell dimensions and symmetries (*38, 52, 53*), the AASBU method assumes only 3-dimensional periodicity; no constraint on the unit cell shape or dimensions is imposed, other than that required by the selected space group symmetry (if symmetry other than triclinic, P1, is defined). Although traditional zeolite chemistry is not, in general, a good example, there are systems in which particular building units are believed to participate in nucleation and growth from particular hydrothermal crystallization media, with the nature of the crystalline product, that is, the manner in which the building units are assembled in 3-dimensions, being influenced by the synthesis conditions (*51*). In such cases, the AASBU method can prove a useful means of articulating the structures which are possible, with the nature of each such structure conceivably then helping in selection of suitable synthesis composition and conditions.

Heterogeneous Catalyst Experimental Design

The coupling between molecular-level design and experimental design in the heterogeneous catalysis field is today still tenuous. Design of experiment ("DOE") tools support the choice of which experimental points to sample in a complex parameter space. Full coverage of the parameter space defined by just the compositional dimensions of a multi-element inorganic system would require an infinite number of experiments. Thus the practitioner need decide (i) how many experiments to perform (depending in part of the accessible experimental throughput); (ii) at what increments each variable is sampled. Based on assumptions about the nature of the experimental space, DOE tools suggest the best coordinates for measurement. These assumptions include, implicitly or explicitly, the smoothness of the variation in measured property with change in each of the variables. If the assumptions are valid, the set of measurements proposed by the DOE tools will then be sufficient to characterize the parameter space. The activity, or selectivity, (or some combination of the two) of a given heterogeneous catalyst system might vary smoothly with temperature or mixing rate, but vary dramatically, and perhaps discontinuously, with composition or pretreatment conditions. Thus DOE tools provide but one aid in the design process, with a value that can increase as understanding of the parameter space is accumulated in successive iterations through the HTE cycle (Figure 2).

Descriptor-Property Relationships (DPR) and Quantitative Descriptor-Property Relationships (qDPR)

Several technologies are being developed to contribute to the Design stage (Figure 2) (7, 19). For the Model stage, as captured in a systems view (19-22), it is ultimately structure that determines properties. This has been the basis for the field directed to quantitative structure-property relationships (QSPR) or quantitative structure-activity relationships (QSAR) (for some example applications in the soft materials field see, for example, (54-58)).

The vast majority of atomic-level attributes that are quantified and considered as potential governors of properties by these methods, however, are not direct structural data, but rather 'descriptors', that is, attributes that depend on the structure. Descriptors for molecular QSAR or QSPR, many of which can be computed readily from the structural formula (2-Dimensional) or the molecular structure (3-Dimensional), might capture aspects of the molecular topology, structure, conformational space, flexibility, structural fragments, surface, spatial, thermodynamic or electronic structure etc. and be computed by any of the various methods of molecular simulation (*37*). More that 200 descriptor types have been explored, with greater or lesser degress of general utility; topological indices have, for example, been well validated for many classes of polymers (*59*).

If descriptors are suitably chosen, then different materials that have similar values of the descriptors will have similar target properties. If a library of compounds is represented as points in the N-dimensional descriptor space, or in the space of its principal components, compounds that are close together will have similar property values and compounds that are far apart will have dissimilar property values. Thus library design, visualization and analysis in the Design stage is also preferably applied referenced to the descriptors values.

The relationships developed in the Model stage are formally between descriptor values and properties, not between 'structure' and properties. Further, in heterogeneous catalysis we rarely have the benefit of definitive structural data at the atomic level and the descriptors in use are then often one step further removed from such structure. Nonetheless, it is practicable to consider qualitative descriptor-property relationships ('activity' is considered a property) (DPR) and quantitative descriptor-property relationships (qDPR) in many cases. Given that a useful inferential and correlative infrastructure is already implemented for molecular applications (see, e.g., (*19*) for references), the prime challenge in heterogeneous catalysis is the development and validation of suitable descriptors.

Issues and Development Opportunities

As with many of the experimental aspects of high throughput experimentation, the computational components are undergoing rapid change. Although the pace of developments to date has, perhaps arguably, been impressive, there remain ample opportunities for developing more creative ways of addressing, or of improving the approaches thus far developed, to the issues introduced above. Some key design considerations in the development of the types of high capacity MatInformatics systems required include:

- completeness and usability:
    - a preference is that any manual entry of data be minimized

- usability is detemined in significant part by the ease of incorporating and accessing information
- flexibility and extensibility:
  - the useful half-life on many experimental components, and even the overall approach to experimentation, is relatively short
- transferability:
  - the requirements of one HTE program are rarely identical to the next, even within one sub-field such as heterogeneous catalysis
- compatibility
  - with other systems, recent or historical
  - with other internal and external information sources, open and patent literature, accessible materials information databases
  - with internal and third party tools for data and structure visualization, statistical packages, emerging new methods for Design and Model stages
  - with diverse experimental components, such as analytical instruments, used locally or remotely to accumulate data subsets

As with any data base system, the value of the system is heavily dependent on data quality. Decisions then need to be made relative to the degree that historical data might be included, or to the extent that experimental areas already explored might be re-visited with a suitable HTE system, so as to populate the data base with consistent and reliable data.

The software engineering issues are by no means trivial, but while the subject matter considered here is HTE-specific, similar informatics engineering issues are being explored, in parallel, in many companion fields. Research on methods that might impact on the Design and Model stages is also pursued by several related disciplines.

Conclusion

It is practicable in applying Stage 2 HTE technology to heterogeneous catalysis to achieve a caliber of data in the assesment of catalytic activity and selectivity that compares reasonably with conventional, laboratory-scale experimentation (*16*). Although published data are yet sparse, the indications are also that the automation of traditional catalyst preparation procedures can not simply match historical manual operations, but yield improved consistency and reproducibility (*16*). We can then argue that the results from a single catalyst evaluation in a parallel Stage 2 HTE experiment are as informative as those from a conventional experiment. The MatInformatics environment, however, provides a major additional value, in that a datum can contribute not only to progress in the immediate project, but be part of an expanding and accessible information repository that can be tapped, in even as yet unanticipated fashions, on an ongoing basis into the future. It may, as a result, be argued, that perhaps 80% of the value of today's HTE approaches will arise not from the degree of parallelism in reactor technology, the speed of robotic devices or other types of hardware improvements, but in the experimental and simulation data themselves and how their value is mined.

Acknowledgements

We thank many colleagues for providing materials in advance of publication and for numerous ideas and suggestions. MSI's Catalysis and Sorption, and Combinatorial Chemistry projects are each supported by a consortium of industrial, academic and government institutions; we thank the memberships for their guidance, input and for stimulating discussions. We also acknowledge contributions to the work outlined here by our many hte colleagues and collaborators, most especially J. Baldwin, A. Brenner, R. Brown, M. Doyle, B. E. Eichinger, C. M. Freeman, D. King-Smith, S. Schunk, F. Schüth and W. Stichert.

References

1.   Gennari, C.; Nestler, H. P.; Piarulli, U.; Salom, B. Combinatorial Libraries: Studies in Molecular Recognition and the Quest for New Catalysts. *Liebigs Ann. Recueil* **1997**, 637-647.

2.   Shimizu, K. D.; Snapper, M. L.; Hoveyda, A. H. High-Throughput Strategies for the Discovery of Catalysts. *Chem. Eur. J.* **1998**, *4*, 1885-1889.

3.   Jandeleit, B.; Turner, H. W.; Uno, T.; van Beek, J. A. M.; Weinberg, W. H. Combinatorial methods in catalysis. *Cattech* **1998**, *2*, 101-123.

4.   Francis, M. B.; Jamison, T. F.; Jacobsen, E. N. Combinatorial libraries of transition-metal complexes, catalysts and materials. *Curr. Opin. Chem. Biol.* **1998**, *2*, 422-428.

5.   Bein, T. Efficient Assays for Combinatorial Methods for the Discovery of Catalysts. *Angew. Chem. Int. Ed.* **1999**, *38*, 323-326.

6.   Newsam, J. M.; Schüth, F. Combinatorial approaches as a component of high throughput experimentation (HTE) in catalysis research. *Biotechnology and Bioengineering (Combinatorial Chemistry)* **1999**, *61*, 203-216.

7. *Combinatorial Catalysis and High Throughput Catalyst Design and Testing (NATO Science Series C: Vol. 560)*; Derouane, E. G.; Lemos, F.; Corma, A.; Ribeiro, F. R., Eds. Kluwer Academic Publishers: Dordrecht, Netherlands, 2000.

8. Deem, M. W. A Statistical Mechanical Approach to Combinatorial Chemistry. *Chemical Engineering* **2000**, in press.

9. Falcioni, M.; Deem, M. W. Library Design in Combinatorial Chemistry by Monte Carlo Methods. *Phys. Rev. E.* **2000**, *61*, 5948-5952.

10. Moates, F. C.; Somani, M.; Annamalai, J.; Richardson, J. T.; Luss, D.; Willson, R. C. Infrared thermographic screening of combinatorial libraries of heterogeneous catalysts. *Ind. Eng. Chem. Res.* **1996**, *35*, 4801-4803.

11. Taylor, S. J.; Morken, J. P. Thermographic Selection of Effective Catalysts from an Encoded Polymer-Bound Library. *Science* **1998**, *280*, 267-270.

12. Holzwarth, A.; Schmidt, H.-W.; Maier, W. F. Detection of catalytic activity in combinatorial libraries of heterogeneous catalysts by IR thermography. *Angew. Chemie Int. Ed.* **1998**, *37*, 2644-2647.

13. Senkan, S.; Krantz, K.; Ozturk, S.; Zengin, V.; Onal, I. High-Throughput Testing of Heterogeneous Libraries Using Array Microreactors and Mass Spectrometry. *Angew. Chem. Int. Ed.* **1999**, *38*, 2794-2799.

14. Orschel, M.; Klein, J.; Schmidt, H.-W.; Maier, W. F. Detection of Reaction Selectivity on Catalyst Libraries by Spatially Resolved Mass Spectrometry. *Angew. Chem. Int. Ed.* **1998**, *38*, 2791-2794.

15. Cong, P.; Doolen, R. D.; Fan, Q.; Giaquinta, D. M.; Guan, S.; McFarland, E. W.; Poojary, D. M.; Self, K.; Turner, H. W.; Weinberg, W. H. High-Throughput Synthesis and Screening of Combinatorial Heterogeneous Catalyst Libraries. *Angew. Chem. Int. Ed.* **1999**, *38*, 484-488.

16. Hoffmann, C.; Wolf, A.; Schüth, F. Parallel synthesis and testing of catalysts under nearly conventional testing conditions. *Angew. Chem. Int. Ed.* **1999**, *38*, 2800-2803.

17. Senkan, S. M. High-throughput screening of solid-state catalyst libraries. *Nature* **1998**, *394*, 350-353.

18. Senkan, S. M.; Ozturk, S. Discovery and Optimization of Heterogeneous Catalysts Using Combinatorial Chemistry. *Angew. Chem. Int. Ed.* **1999**, *38*, 791-795.

19. Newsam, J. M. Design of Catalysts and Catalyst Libraries, In *Combinatorial Catalysis and High Throughput Catalyst Design and Testing (NATO Science Series C: Vol. 560)*, Derouane, E. G.; Lemos,

F.; Corma, A.; Ribeiro, F. R., Eds., Kluwer Academic Publishers: Dordrecht, Netherlands, 2000; p. 301-335.

20. Newsam, J. M.; Li, Y. S. A multi-faceted approach to modeling heterogeneous catalysts. *Catalysis Today* **1995**, *23*, 325-332.

21. Misono, M. New catalytic aspects of heteropoly acids and related compounds- to the molecular design of practical catalysts, In *Stud. Surf. Sci. Catal. Vol. 75 (New Frontiers in Catalysis, Part A)*, Elsevier Science B.V.: Amsterdam, 1993; p. 69-101.

22. Misono, M. Catalytic reduction of nitrogen oxides by bifunctional catalysts. *Cattech* **1998**, *2*, 183-196.

23. Sie, S. T.; Krishna, R. Process development and scale up: II. Catalyst design strategy. *Rev. Chem. Eng.* **1998**, *14*, 159-202.

24. Becker, E. R.; Pereira, C. J. *Computer Aided Design of Catalysts*, Marcel Dekker, Inc.: New York, 1993.

25. Thomas, J. M. Solid state chemistry and the design of heterogeneous catalysts (1950-1999). *Catal. Lett.* **2000**, *67*, 53-59.

26. Dusi, M.; Mallat, T.; Baiker, A. Epoxidation of functionalized olefins over solid catalysts. *Catal. Rev. - Sci. Eng.* **2000**, *42*, 213-278.

27. Ravasio, N. Use of heterogeneous copper catalysts and of acidic mixed oxides in organic synthesis. *Recent Res. Dev. Org. Chem.* **1999**, *3*, 79-85.

28. Coq, B. Metal-support interaction in catalysis: generalities, basic concepts and some examples in hydrogenation and hydrogenolysis reactions, In *Metal-Ligand Interactions in Chemistry, Physics and Biology (NATO Science Series C: Volume 546)*, Corma, A.; Melo, F. V.; Mendioroz, S.; Fierro, J. L. G., Eds., Kluwer Academic Publishers: Dordrecht, Netherlands, 2000; p. 49-71.

29. Centi, G.; Perathoner, S. Oxidation catalysts: new trends. *Curr. Opin. Solid State Mater. Sci.* **1999**, *4*, 74-79.

30. Besenbacher, F.; Chorkendorff, I.; Clausen, B. S.; Hammer, B.; Molenbroek, A. M.; Nørskov, J. K.; Stensgaard, I. Design of a Surface Alloy Catalyst for Steam Reforming. *Science* **1998**, *279*, 1913-1915.

31. McLeod, A. S.; Gladden, L. F. Heterogeneous Catalyst Design Using Stochastic Optimization Algorithms. *J. Chem. Inf. Comput. Sci.* **2000**, *40*, 981-987.

32. Hutchings, G. J.; Willock, D. J. Heterogeneous enantioselective catalysts: can molecular simulation techniques aid the design of improved catalysts? *Top. Catal.* **1998**, *5*, 177-185.

33. Hutchings, G. J.; Bethell, D.; McGuire, N.; Page, P. C. B.; Robinson, D.; Willock, D. J.; Hancock, F.; King, F. Heterogeneous catalysts by design. *Curr. Top. Catal.* **1999**, *2*, 39-58.

34.   Choplin, A.; Coutant, B.; Dubuisson, C.; Leyrit, P.; Mcgill, C.; Quignard, F.; Teissier, R. Heterogeneous catalysts from organometallic precursors: how to design isolated, stable and active sites. Applications to zirconium catalyzed organic reactions, In *Stud. Surf. Sci. Catal. Volume 108 (Heterogeneous Catalysis and Fine Chemicals IV)*, Blaser, H. U.; Baiker, A.; Prins, R., Eds., Elsevier Science B.V.: Amsterdam, 1997; p. 353-360.

35.   Herron, N.; Farneth, W. E. The design and synthesis of heterogeneous catalyst systems. *Adv. Mater.* **1996**, *8*, 959-968.

36.   Davis, M. E. Molecular design of heterogeneous catalysts, In *Stud. Surf. Sci. Catal. Volume 130A (International Congress on Catalysis, 2000, Pt. A)*, Corma, A.; Melo, F. V.; Mendioroz, S.; Fierro, J. L. G., Eds., Elsevier Science B.V.: Amsterdam, 2000; p. 49-59.

37.   Andzelm, J. W.; Alvarado-Swaisgood, A. E.; Axe, F. U.; Doyle, M. W.; Fitzgerald, G.; Freeman, C. M.; Gorman, A. M.; Hill, J.-R.; Kölmel, C. M.; Levine, S. M.; Saxe, P. W.; Stark, K.; Subramanian, L.; van Daelen, M. A.; Wimmer, E.; Newsam, J. M. Heterogeneous Catalysis: Looking Forward with Molecular Simulation. *Catalysis Today* **1999**, *50*, 451-477   (see   also http://www.msi.com/materials/articles/catavis/CATAVIS.HTM).

38.   Deem, M. W.; Newsam, J. M. Framework Crystal Structure Solution by Simulated Annealing. Test Application to Known Zeolite Structures. *J. Amer. Chem. Soc.* **1992**, *114*, 7189-7198.

39.   Santilli, D. S.; Harris, T. V.; Zones, S. I. Inverse shape selectivity in molecular sieves: Observations, Modeling and Predictions. *Microporous Materials* **1993**, *1*, 329-341.

40.   Ricchiardi, G.; Newsam, J. M. The Predicted Effects of Site-Specific Aluminum Substitution on the Framework Geometry and Unit Cell Dimensions of Zeolite ZSM-5 Materials. *J. Phys. Chem.* **1997**, *101B*, 9943-9950.

41.   Newsam, J. M.; Freeman, C. M.; Gorman, A. M.; Vessal, B. Simulating Non-Framework Cation Location in Aluminosilicate Zeolites. *J. Chem. Soc. Chem. Comm.* **1996**, 1945-1946.

42.   Gorman, A. M.; Kölmel, C. M.; Freeman, C. M.; Newsam, J. M. Accelerated approach to non-framework cation placement in crystalline materials. *Faraday Discussions* **1997**, *106*, 489-494.

43.   Lignières, J.; Newsam, J. M. Simulations of the Non-framework Cation Configurations in Dehydrated Na-Ca and Na-Li Zeolite A. *Microporous and Mesoporous Materials* **1999**, *28*, 305-314.

44.   Toulhoat, H.; Raybaud, P.; Kasztelan, S.; Kresse, G.; Hafner, J. Transition metals to sulfur binding energies relationship to catalytic

activities in HDS: back to Sabatier with first principle calculations. *Catalysis Today* **1999**, *50*, 629-636.

45. Koy, J.; Ladebeck, J.; Hill, J.-R. Role of Cr in Fe based High Temperature Shift Catalysts, In *Natural Gas Conversion V (Studies in Surface Science and Catalysis Vol. 119)*, Parmaliana, A., Ed., Elsevier Science B.V.: Amsterdam, 1998; p. 479-484.

46. Christensen, A.; Carter, E. A. First-Principles Study of the Surfaces of Zirconia. *Phys. Rev.* **1998**, *B 58*, 8050-8064.

47. Wang, X.-G.; Weiss, W.; Shaikhutdinov, S. K.; Ritter, M.; Petersen, M.; Wagner, F.; Schlogl, R.; Scheffler, M. The hematite (alpha-Fe2O3) (0001) surface: Evidence for domains of distinct chemistry. *Phys. Rev. Lett.* **1998**, *81*, 1038-1041.

48. van Daelen, M. A.; Li, Y. S.; Newsam, J. M.; van Santen, R. A. Energetics and Dynamics for NO and CO Dissociation on Cu(100) and Cu(111). *J. Phys. Chem.* **1996**, *100*, 2279-2289.

49. Ge, Q.; King, D. A. The Chemisorption and Dissociation of Ethylene on Pt{111}. *J. Chem. Phys.* **1999**, *110*, 4699-4702.

50. Yajima, K.; Ueda, Y.; Tsuruya, H.; Kanougi, T.; Oumi, Y.; Ammal, S. S. C.; Takami, S.; Kubo, M.; Miyamoto, A. Computer-aided design of novel heterogeneous catalysts - a combinatorial computational chemistry approach, In *Stud. Surf. Sci. Catal. Volume 130A (International Congress on Catalysis, 2000, Pt. A)*, Corma, A.; Melo, F. V.; Mendioroz, S.; Fierro, J. L. G., Eds., Elsevier Science B.V.: Amsterdam, 2000; p.

51. Mellot Draznieks, C.; Newsam, J. M.; Gorman, A. M.; Freeman, C. M.; Férey, G. De Novo Prediction of Inorganic Structures Developed through Automated Assembly of Secondary Building Units (AASBU) Method. *Angew. Chem. Int. Ed.* **2000**, *39*, 2270-2275.

52. Deem, M. W.; Newsam, J. M. Determination of 4-Connected Framework Crystal Structures by Simulated Annealing. *Nature* **1989**, *342*, 260-262.

53. Falcioni, M.; Deem, M. W. A biased Monte Carlo scheme for zeolite structure solution. *J. Chem. Phys.* **1999**, *110*, 1754-1766.

54. Livingstone, D. *Data Analysis for Chemists. Applications to QSAR and Chemical Product Design*, Oxford University Press: Oxford, 1995.

55. Johnson, S. R.; Jurs, P. C. Prediction of the Clearing Temperatures of a Series of Liquid Crystals from Molecular Structure. *Chem. Mater.* **1999**, *11*, 1007-1023.

56. Patel, H. C.; Tokarski, J. S.; Hopfinger, A. J. Molecular Modeling of Polymers. 16. Gaseous Diffusion in Polymers: a Quantitative Structure-Property Relationship (QSPR) Analysis. *Pharmaceutical Research* **1997**, *14*, 1349-1354.

57.   Sabljic, A. Calculation of retention indexes by molecular topology. Chlorinated benzene. *J. Chromatogr.* **1985,** *319,* 1-8.

58.   Yao, S.; Shoji, T.; Iwamoto, Y.; Kamei, E. Consideration of an activity of the metallocene catalyst by using molecular mechanics, molecular dynamics and QSAR. *Computational and Theoretical Polymer Science* **1999,** *9,* 41-46.

59.   Bicerano, J. *Prediction of Polymer Properties*, Marcel Decker, 1993.

Chapter 9

Epilogue

Repudaman Malhotra

SRI International, 333 Ravenswood Avenue, Menlo Park, CA 94025

As I get ready to submit all the chapters of this book to the publisher, I realize that eighteen months have elapsed since the symposium in which these talks were delivered. For a rapidly growing technology, that is a significant amount of time, and hence it may be worthwhile to take a look back at the progress in this field since the symposium. A caveat: This epilogue is a personal account based on the impressions I have received through the normal course of my "day job." In other words, this essay is not a research review but a collection of items that caught my fancy while I was reading journals and news magazines, attending meetings, and holding informal conversations with colleagues. Reports by Stu Borman[1] and Ron Dagani[2] in the *Chemical and Engineering News* have helped me keep up with the field of combinatorial chemistry. Some of the information in this essay was gleaned from the press releases posted on the web sites of companies such as Symyx Technologies, Argonaut Technologies, Avantium Technologies, HTE, and Millenium Cell.

From an economic perspective, eighteen months ago was quite a different time. The U.S. economy was enjoying a long sustained growth and the markets were in the midst of an "irrational exuberance." A serious downturn has occurred since then, and with it a deflating of the markets. The reason for bringing up changes in economic conditions in this technical book is that significant research on combinatorial approaches for materials development is being conducted in small companies that depend on the market for funding their R&D activities. The substantial erosion of the market capitalization of these companies has undoubtedly slowed down progress.

To the best of my knowledge, no electronic, luminescent, magnetic, or catalytic material developed by combinatorial methodology is currently in commercial production. Perhaps the closest to being marketed is the X-ray storage phosphor developed by Symyx Technologies under a contract with Agfa[2b],[3] There are also many tantalizing news releases of new polymerization catalysts, but no commercial product has been announced. It is perhaps too early to expect a commercial product. After all, even after the discovery of a catalyst or a new material, it can easily take three to five years of development before the products make it to the market. Commercial products aside, significant advances can be found both in the methods of practicing and the materials developed by using combinatorial approaches.

Methods

Three developments in the basic strategy of combinatorial chemistry deserve mention. These are: (i) continuous composition spread (CCS), also known as continuous phase diagram (CPD); (ii) dynamic combinatorial libraries (DCL); and (iii) dual recursive deconvolution (DRED). Adoption of these methods is increasing, and each of them greatly facilitates the discovery process.

Most of the earlier work on combinatorial syntheses centered on making compounds in individual reaction vessels or spots. This method was particularly suited for liquids and for preparing discrete structures and testing them for pharmacological activity. The continuous composition spread methodology, described in Chapters 2 and 3 of this book, relies on forming compositional gradients of the constituents. Many researchers are now applying the CCS method. When coupled with scanning microprobe techniques for determining the composition and properties of the resultant phases, this method can be used to test entire phase diagrams of binary and ternary mixtures. Some examples of the use of CCS are discussed below.

A second innovation is the adoption of the methods from the study of self-assembly of supramolecular systems. In this approach a number of molecules are allowed to react with one another to form an equilibrium mixture of complexes or a dynamic combinatorial library (DCL).[4] When a species that binds with the complexes is introduced into the mixture, the equilibrium is shifted toward those specific complexes that are most reactive to the probe. The technique can be used to build either assemblies that would fit in a target receptor or ones that would encase the target. Sanders and coworkers have used the reaction of hydrazones with ketones in forming libraries containing linear and cyclic versions of dimeric, trimeric, and higher pseudopeptides.[5] When they added 18-crown-6 to this mixture, the equilibrium shifted to form the specific hydrazone macrocycle that has the highest affinity for 18-crown-6. Thus, not only the system self-screened itself, it also used the pool of reactants to selectively express the desired product. Initial applications of the DCL method have been in the area of drug discovery; however, this approach is also well suited to the discovery of functional materials such as sensors, selective membranes, and stationary phases for chromatography.

The first two innovations discussed above deal with the synthesis of materials. The third one—Dual Recursive Deconvolution (DRED)—deals with the process of deconvoluting the composition of the active material after it passes the screening process.[6] This approach, developed by Fenniri and coworkers, is useful when libraries are synthesized by a sequence of many reactions on beads by the pool and split method. After a target molecule has been found, the last step in the sequence of its synthesis is identified from the pool screening. The process is repeated to recursively identify the entire history of steps that led to its synthesis—each synthesis step requiring one recursion.

The DRED method consists of first developing a set of resin beads that are coded with unique optical signatures. The first building blocks of the combinatorial sequence are then bonded to beads of a given optical signature. As before, at the end of the multistep synthesis and screening, the identity of the last synthesis step can be inferred from the last pool in which the target molecule was found. Simultaneously, the first step in this process can be inferred from the optical signature of the bead. Thus, during deconvolution two synthesis steps are identified. By removing these steps from the sequence and iterating the procedure with the remaining steps, the identity of the active compound can be deciphered at a much faster pace. The DRED method is particularly advantageous when the pool of target molecules is very large, as it can be ten times faster than the fastest of the more traditional approaches. Fenniri and his colleagues recently reported on the development of 24 polymeric resins by copolymerizing styrene with various analogs.[7] The resulting beads have unique IR and Raman signatures, or barcodes, that are not affected by the subsequent reactions used in attaching and modifying the target molecules.

Materials

As mentioned above, the use of CCS methodology for materials discovery has been gaining popularity. Xiang and coworkers have developed an *in situ* shutter system in a pulsed laser deposition chamber with eight targets that can be used to prepare films of uniform thickness over a 15 mm by 15 mm square. The two shutters, one moving horizontally and the other vertically, allow for varying the amounts of the constituents deposited across the film. These researchers reported on the preparation and electronic properties of perovskite manganites of the general formula $RE_{1-x}A_xMnO_3$, where RE is one of the rare earth metals Eu, Gd, Tb, Er, Tm, and Yb; A is Ca, Sr, or Ba; and x is continuously varied from 0 to 1. The as-deposited films were annealed at temperatures up to 1000°C. The high-temperature treatment led to the development of distinct crystalline phases. This study showed that there are several phases with extremely narrow compositional variation with very different electronic properties, including some singular phases around x = 7/8 in the erbium calcium manganite on $NdGaO_3$ substrate and yitterbium calcium manganite on $SrTiO_3$ substrate. Such phases could easily be missed in studies that depend on preparing discrete compositions.

As reported by Dagani, Ramirez and others at Agere are fabricating new alloys that would combine good electrical conductivity with good mechanical properties.[2] The goal is to find materials suitable for application in high-speed micromechanical switches. Gold is a very good conductor and is durable because it resists corrosion; however, it lacks the necessary mechanical stiffness. The researchers at Agree have accordingly made CCS libraries of gold with metals such as antimony and cobalt in their search for the desired alloy. One interesting feature of this research was the finding that many of the as-prepared compositions showed considerably reduced conductivity, but upon annealing, they regained some of the lost conductivity. This observation serves as another reminder that materials prepared in micro-scale using vapor deposition techniques can turn out to be quite different when prepared in bulk.

Continuous composition libraries can be grown in bulk by contacting different metals and heating them. This approach has been used by Zhao to prepare and investigate the phase diagram of structural (load bearing) materials, including superalloys used in hot engine parts.[8] For example, the Ni-NiAl phase diagram was investigated by joining pieces of Ni and NiAl and heating the couple under hydrostatic pressure at 1200°C for extended periods to facilitate interdiffusion. The couple thus produced was sectioned, polished, and examined. The composition of the resulting phases was determined by using electron probe microanalysis, and the mechanical properties were determined by using a nanoindentor probe. The nanoindentor measures the extent of penetration of a diamond tip at various loads, and from these data hardness and moduli can be determined with high spatial resolution. For the purpose of making ternary and quaternary systems, the individual metals are cut in the shape of sectors of a cylinder and subjected to thermal treatment under hydrostatic pressure. The region near the center, where all the metals meet, is where the interdiffusion occurs. By studying the Fe-Ni-Mo system, Zhao was able to provide clear evidence for the role of molybdenum in imparting hardness to the alloy. In a *News and Views* piece about this work for *Nature*, Cahn points out that whereas both thin-film and interdiffusion approaches are suitable for determining the phase diagrams, only the interdiffusion method is suitable for assessing the mechanical properties of the resulting materials.[9]

A recent report in *Science* on the discovery of room-temperature ferromagnetic semiconductor films by Koinuma and his coworkers provides a very interesting example of how combinatorial libraries can lead to advancements in multiple arenas.[10] Prompted by theoretical work that suggested doping of zinc oxide with 3-d transition metals could lead to dilute ferromagnets, Koinuma's group had prepared a library of such compounds by laser-depositing different 3-d elements on a ZnO film. However, none of the compositions showed the desired ferromagnetism. All was not lost, because the deposition techniques developed were used by another part of his team engaged in the development of catalysts for oxidizing hazardous wastes. This team prepared and tested a library of titania doped with various 3-d transition metals. Having prepared the TiO_2 library, they decided to see if any of these compositions exhibited ferromagnetism. This time, the effort was successful. By using a scanning SQUID magnetometer they identified titania doped with up to 8% cobalt as being ferromagnetic. Doping with other transition metals did not produce ferromagnetism. The discovery of ferromagnetic semiconductors is critical to the development of spintronic devices, and in a *Perspective* accompanying the paper by Koinuma, Ohono writes of the possible applications of this material ranging from "refrigerator magnets to mass storage in information technology."[11] He also points out that the physical basis of the observed ferromagnetism in these films is not clear. The importance of this finding is also evidenced by the efforts other groups have launched to build on Koinuma's work. In his recent survey, Dagani reports that a collaboration between researchers at IBM's Almaden Laboratories and the Pacific Northwest National laboratory is investigating cobalt-doped TiO_2 films prepared by oxygen-plasma-assisted molecular beam epitaxy.[2] This method produces nearly defect-free films. This group has ruled out inclusion of metallic cobalt in the oxide film as the source of ferromagnetism, and moreover has shown that the ferromagnetism is stable up to 250°C.

Braithwaite and Barbour at Sandia National Laboratory have applied the combinatorial approach to study the corrosion of copper interconnects in semiconductor devices.[12] They deposited Cu on a silicon wafer and used lithography to prepare thin lines. Various dopants were ion-implanted in the lines and the wafer was exposed to low levels of hydrogen sulfide. The formation of copper sulfide was monitored by the changes in line thickness and electrical resistivity. These tests showed that indium had a marked effect in retarding the corrosion of copper lines.

All of the above examples deal with the development of inorganic materials. A recent report on the use of combinatorial techniques for improving the performance of organic light-emitting diodes (OLEDs) provides a counterpoint. OLEDs are generally fabricated by sandwiching a film of π-conjugated polymers between electrodes. A transparent film of indium tin oxide (ITO) is often used as the anode. By applying an electric field, the electrons from the polymer are removed by the anode creating holes, while electrons are injected to the polymer near the cathode. Diffusion of the charged organic species and neutralization of the charges leads to the formation of excitons, which then decay by giving off light. Because the work function of ITO is generally less than the energy required to ionize the polymer, there is a barrier to hole injection. In ITO the hole barrier can be adjusted by doping with impurities. Gross et al. have investigated the use of conductive organic polymers poly(3,4-ethylenedioxythiophene) (PEDOT) and poly(4.4'-dimethoxy-bithiophene) (PDBT) as anodes for OLED application.[13] They reasoned that the hole barrier in these systems could be adjusted by controlled electrochemical oxidation. Furthermore, since the transport properties can also be affected by the film thickness, this parameter too must be controlled. The authors report a combinatorial experiment in which the electroluminescence was investigated as a function of redox level and film thickness of the polymer. The application of combinatorial methods appears eminently suitable for formulating polymer compositions for different applications.

Catalysis. Catalysis remains a major focus of combinatorial materials research. Press releases from Symyx Technologies indicate that several full-scale systems have been sold to large corporations such as ExxonMobil and Dow Chemicals. Symyx has also partnered with Argonaut Technologies to sell its Endeavor system. Avantium Technologies, which was formed as a spin-off from Shell in February 2000, is now collectively owned by several chemical, petrochemical, and pharmaceutical companies and three of Holland's universities, and has brought together broad expertise in combinatorial chemistry. Avantium offers tools for catalyst development and also engages in the R&D of catalysts. In a collaboration with Millenium Cell, Avantium has developed a catalyst for use in a hydrogen-on-demand system for fuel cells. Another major player in this arena is HTE, which also offers a complete package of tools for high-throughput experimentation and conducts R&D in collaboration with other companies such as BASF and Chevron. Besides these major players, many other companies, such as TDA Reseach, Parr, Autoclave Engineers, and Zeton Altamira, have also entered this field and are offering systems for the discovery and study of catalysts. A broad patent on combinatorial catalysis was recently issued to Symyx, but how this will affect the market structure remains to be seen.

High-throughput screening of catalysts presents formidable challenges. Techniques commonly employed for parallel or rapid sequential analysis include IR-thermography, acid-base indicators, gas chromatography, and mass spectrometry. Previously, Senkan had reported on a system for high-throughput screening of heterogeneous catalysts. He described a 72-site system that used a resonance-enhanced multiphoton ionization (REMPI) spectroscopy with a microelectrode detector for monitoring the products.[14] Recently, Senkan and Ozturk reported the dehydrogenation of cyclohexane to benzene over an array of microreactors containing various combinations of Pt-Pd-In.[15] By the REMPI method they were able to ascertain that an alloy containing 80% Pt with 10% each of Pd and In gave the highest conversion under their conditions. The authors also mention in this paper that the total time for preparing and calcining the 85 catalysts and then screening them in five sequential tests of 17 catalysts was only 2.5 days!

In catalysis research, combinatorial methods are increasingly being used for optimizing process conditions. The availability of tools that can handle multiple reactors and analytical trains in parallel has also meant that reactions with the same substrate and catalysts can be conducted under different conditions of temperature, pressure, or other variables simultaneously. To some extent such work has been conducted previously by using equipment such as ShakerClave by Autoclave Engineers that accommodate multiple high-pressure reactors with independent gas lines, albeit at the same temperature. The new tools have extended that capability dramatically affording a much higher degree of flexibility in the number of reactors and the parameters that can be independently varied.

From the foregoing, it would appear that combinatorial discovery and development of materials is poised for rapid growth. Advances in miniaturization and the integration of informatics are surely going to facilitate wider adoption of combinatorial methods. Application of this technology has led to the discovery of new materials as well as to the optimization of processes. The jury is still out as to whether it will live up to its promise of increased productivity, innovation, and shortened R&D cycles.

References

[1] a) Borman, S. *Chemical and Engineering News* **1999**, *77*, 33; b) **2001**, *79*, 49.

[2] a) Dagani, R. *Chemical and Engineering News* **2000**, *78*, 66; b) **2001**, *79*, 59.

[3] Symyx Technologies, *Product Pipeline* www.symyx.com.

[4] Lehn, J.-M.; Eliseev, A. V. *Science* **2001**, *291*, 2331.

[5] Sanders, Chem. Commun. **2000**, 1761.

[6] Fenniri, H.; Hhedderich, H. G.; Haber, K. S.; Achkar, J.; Taylor, B.; Ben-Amotz, D. *Angew. Chem. Intl Ed.* **2000**, *39, 4483*.

[7] Fenniri, H.; Ding, L.; Ribbe, A. E.; Zyrianov, Y. *J. Am. Chem. Soc.* **2001**,

[8] Zhao, J.-C. *Advanced Electronic Materials* **2001**, *3*, 143.

[9] Cahn, R. W., *Nature* **2001**, *410*, 643.

[10] Matsumoto, Y.; Murakami, M.; Shono, T.; Hasegawa, T.; Fukumura, T.; Kawasaki, M.; Ahmet, P.; Chikyow, T.; Koshihara, S.; Koinuma, H. *Science* **2001**, *291*, 854.

[11] Ohno, H. *Science* **2001**, *291*, 840.

[12] Barbour, J. C.; Braithwaite, J. W.; Wright, A. F. *Nucl. Instrum. Methods, Phys. Res. B* **2001**, *175-177*, 382-387.

[13] Gross, M.; Müller, D. C.; Nothofer, H.-G; Scherf, U.; Neher, D.; Bräuchle, C.; Meerholz, K. *Nature* **2000**, *405*, 661.

[14] Senkan, S. M. *Nature* **1998**, *394*, 351.

[15] Senkan, S. M.; Ozturk, S. *Angew. Chem. Int. Ed.* **1999**, *38*, 791.

INDEXES

Author Index

Subject Index

A

Acceleration, catalyst screening study, 87-106

Alcohol oxidation to aldehyde study, 78-83

Amino acids used in peptide sequence generation, 111t

Amorphous microporous mixed oxide materials, reaction for silica based, 8

Amorphous microporous mixed oxides library, 7-11

Arrayed waveguide grating routers, 58-60
See also Planar lightwave circuits

Automated data reduction in experiment understanding, 134-135

B

Beads carrying different ligands, identification in resin supported libraries, 76-77f
See also Resin bead supported libraries

Biodegradable polyarylates, combinatorial library preparation, 26-27

Block copolymer segregation, symmetric diblock copolymer thin films, 42-43

C

Calculation methods, sequence-conformation probabilities, 110-117

Capacitor, maximum charge per unit area calculation, 55

Catalysis as focus of combinatorial materials research, 170-171

Catalyst library as array of micro reactors, 12

Catalyst materials in gas phase reactor, 5f-6, 8

Catalyst screening
acceleration study, 87-106
data handling study, 96
quantitative hydrogenation, 96-104
with mass spectrometry, 11-17

Catalysts, polymers, and materials, combinatorial chemistry, 1-21

Catalytic activity, consistency test results, quantitative hydrogenation, 102-103f

Catalytic systems, 153

Characterizing polymers, limitations, 26

Chemical structures
hydrogenation products, 99
materials in parallel hydrogenation, 3-substituted indolin-2-ones, 98f

Chemoselectivity screen, quantitative hydrogenation, 102, 104-105f

Combinatorial chemistry
access to diversity in chemical substances, 2-3
definition, 148-149

Combinatorial data organizational principles, 139-141f

Combinatorial discovery cycle, 129-130f

Combinatorial experimental method, applied to film libraries, schematic, 25f

Bestsellers from ACS Books

The ACS Style Guide: A Manual for Authors and Editors (2nd Edition)
Edited by Janet S. Dodd
470 pp; clothbound ISBN 0–8412–3461–2; paperback ISBN 0–8412–3462–0

Writing the Laboratory Notebook
By Howard M. Kanare
145 pp; clothbound ISBN 0–8412–0906–5; paperback ISBN 0–8412–0933–2

Career Transitions for Chemists
By Dorothy P. Rodmann, Donald D. Bly, Frederick H. Owens, and Anne-Claire Anderson
240 pp; clothbound ISBN 0–8412–3052–8; paperback ISBN 0–8412–3038–2

Chemical Activities (student and teacher editions)
By Christie L. Borgford and Lee R. Summerlin
330 pp; spiralbound ISBN 0–8412–1417–4; teacher edition, ISBN 0–8412–1416–6

Chemical Demonstrations: A Sourcebook for Teachers, Volumes 1 and 2, Second Edition
Volume 1 by Lee R. Summerlin and James L. Ealy, Jr.
198 pp; spiralbound ISBN 0–8412–1481–6
Volume 2 by Lee R. Summerlin, Christie L. Borgford, and Julie B. Ealy
234 pp; spiralbound ISBN 0–8412–1535–9

The Internet: A Guide for Chemists
Edited by Steven M. Bachrach
360 pp; clothbound ISBN 0–8412–3223–7; paperback ISBN 0–8412–3224–5

Laboratory Waste Management: A Guidebook
ACS Task Force on Laboratory Waste Management
250 pp; clothbound ISBN 0–8412–2735–7; paperback ISBN 0–8412–2849–3

Good Laboratory Practice Standards: Applications for Field and Laboratory Studies
Edited by Willa Y. Garner, Maureen S. Barge, and James P. Ussary
571 pp; clothbound ISBN 0–8412–2192–8

For further information contact:
Order Department
Oxford University Press
2001 Evans Road
Cary, NC 27513
Phone: 1-800-445-9714 or 919-677-0977

More Best Sellers from ACS Books

Microwave-Enhanced Chemistry: Fundamentals, Sample Preparation, and Applications
Edited by H. M. (Skip) Kingston and Stephen J. Haswell
800 pp; clothbound ISBN 0–8412–3375–6

Designing Bioactive Molecules: Three-Dimensional Techniques and Applications
Edited by Yvonne Connolly Martin and Peter Willett
352 pp; clothbound ISBN 0–8412–3490–6

Principles of Environmental Toxicology, Second Edition
By Sigmund F. Zakrzewski
352 pp; clothbound ISBN 0–8412–3380–2

Controlled Radical Polymerization
Edited by Krzysztof Matyjaszewski
484 pp; clothbound ISBN 0–8412–3545–7

The Chemistry of Mind-Altering Drugs: History, Pharmacology, and Cultural Context
By Daniel M. Perrine
500 pp; casebound ISBN 0–8412–3253–9

Computational Thermochemistry: Prediction and Estimation of Molecular Thermodynamics
Edited by Karl K. Irikura and David J. Frurip
480 pp; clothbound ISBN 0–8412–3533–3

Organic Coatings for Corrosion Control
Edited by Gordon P. Bierwagen
468 pp; clothbound ISBN 0–8412–3549–X

Polymers in Sensors: Theory and Practice
Edited by Naim Akmal and Arthur M. Usmani
320 pp; clothbound ISBN 0–8412–3550–3

Phytomedicines of Europe: Chemistry and Biological Activity
Edited by Larry D. Lawson and Rudolph Bauer
336 pp; clothbound ISBN 0–8412–3559–7

For further information contact:
Order Department
Oxford University Press
2001 Evans Road
Cary, NC 27513
Phone: 1-800-445-9714 or 919-677-0977